The Wildlife Observer's Eyes

Galen L. Geer

The Wildlife Observer's Eyes

Optical Equipment for Observing Nature

Illustrations by Betty Estes Geer

Printed in the United States of America
Published by Menasha Ridge Press
First edition, first printing

Library of Congress Cataloging-in-Publication Data

ISBN 0–89732–103–0 cloth
ISBN 0–89732–091–3

Cover design by April Leidig-Higgins.
Text design by Bonnie Campbell.

Menasha Ridge Press
Post Office Box 59257
Birmingham, AL 35259-9257

To my mother, Dora E. Geer, who at 80 is still in love with the outdoors.

And to Gail Kimbrel Geer, who manages to keep sharing my love of seeing the outdoors.

Contents

List of Figures

Preface

The world today differs vastly from the one in which I grew up. Computers provide us with new tools every day. Tasks flow faster and smoother. Modern teaching methods allow today's teachers to recognize learning disabilities in youngsters who twenty-five years ago were branded "slow" and became discouraged. Today these youngsters are turned into productive students, often with "discovered" talents and abilities. These students not only keep up with their classmates but they often excel.

But there is a dark side to the modern world. Our thirst for knowledge, understanding, even our simple curiosity about nature is challenged daily by "canned" entertainment. To turn away from a TV show to explore the world in a drop of water through a microscope, to watch a doe through the eyepieces of binoculars, or to gaze at a distant star in a telescope demands a willingness to entertain oneself by thinking.

Today's young adults are masters of computers and modern electronics. Too many have shunned telescopes, binoculars, microscopes, and the other tools of amateur science and observation for overpowered speedboats, status cars, and luxurious homes on acres where sharp-eyed pronghorn once lived. Our world must be seen, heard, and experienced: the early morning flight of geese, a herd of deer appearing from the shadows; tempting a trout from a stream while an eagle cries, and seeing the geometric beauty of a fish scale—all of these experiences enrich our lives.

Our world is alive with beauty. One need not be a wildlife biologist to observe the feeding habits of bighorn sheep. The contributions of amateur observers to science could fill volumes. Knowledge must be a shared experience. Taking the time to look through an eyepiece to see something clearer, closer, or magnified, and sharing that experience, is part of our insurance for the future.

Optics is a science, and the science of optics is better left to those who know and understand the laws of physics involved. Still, within these pages you will be introduced to some laws of science. A basic understanding of these laws will help you understand the optical equipment you now use and will help you make future buying decisions. I hope this book helps you avoid spending too much, and getting too little, in your optical equipment purchases.

I want to give my special thanks to Mr. Don Robertson, product manager, Bushnell/Bausch & Lomb Optics: without Don's help this project would never have been completed. Additional thanks go to Mr. Bob Seamans, Simmons Outdoor Corporation, and Mr. Frank Shadrack, Redfield Optics Corporation.

When you observe nature invite your neighbor and a youngster to share the experience with you: open the doors of our world to others! We live in a wonderful place that needs to be seen, from the heavens to inside a drop of water.

Good observing to you!

The Wildlife Observer's Eyes

1

Why the Looking Glass?

Mysteries invite solutions. My wife, Gail, and I had a mystery in front of us. When we first spotted the cat a few dozen yards from the remote mountain road we thought it was a bobcat. That alone would have been enough to encourage us to grab our binoculars and study the cat. When we did see it through binoculars we realized it was the wrong color, too large, and the ears were not quite right. After studying the animal for several minutes we switched from binoculars to a Bushnell spotting scope and after focusing on the cat we were jubilant to discover we were observing one of the rare tricks of nature, later confirmed by the district wildlife manager. What we had thought at first was a bobcat was confirmed as one of a known pair of Canadian lynxes living in our area. In keeping with the wishes of the wildlife officer the location of the lynx was kept secret and never appeared in any article I wrote.

Everyone who enjoys the outdoors will find their outdoor experience expanded by optical equipment. Whether that equipment is a binocular, spotting scope, telescope, or microscope as every outdoor enthusiast has discovered, optical equipment is an integral part of the outdoor experience; it brings the world of

This young buck heard the "click" of the camera shutter and stopped to look for danger. When he was unable to spot observers he walked away.

wildlife closer without disturbing the animals being observed. Although we spotted the lynx a few dozen yards from a road, we were able to watch it hunt for half-an-hour without disturbing it.

Although today's outdoor enthusiast depends on his optical equipment for observing wildlife in a variety of settings, few users of optical equipment understand what they are using and how much they should spend on their equipment. The "mysteries" of optical equipment confuse even the most confident person into accepting a sales pitch aimed at making a sale rather than providing sound, intelligent advice based on knowledge of the product. Anyone planning to spend their hard-earned cash on optical equipment should get the best equipment for the dollar they spend. This does not mean you should delay buying a pair of binoculars until you save up several hundred dollars for the best on the market. If you have fifty dollars to spend on binoculars buy fifty dollars' worth and not twenty-five dollars' worth of equipment! On the other hand, if you can afford to spend $200 or more, then be sure you are getting your dollar's worth. It is easy to package an inexpensive product in leather and charge three times its real value.

This advice applies to the dedicated bird watcher, sports enthusiast, or amateur naturalist as well as to the professional wildlife biologist. One user's criteria for the purchase of optical equipment can be applied by the others equally. *All* optical equipment for outdoor activities should fall within the following guidelines.

1. The equipment should be worth the dollar paid; that is it should not be overpriced.
2. The equipment should be built to withstand the rough conditions of the outdoors and daily rough handling.
3. The equipment must meet your requirements and needs, not those of the salesman.

Buying equipment to meet those needs is simple: you learn about the product line. The problem is that all too often in a sporting goods store numbers and formulas are whispered like mysterious "words-on-high" passed down to mortal man by an all-knowing god of optics. Don't let yourself be fooled! You can understand the strange lingo and beat the salesman at his own game. That is the purpose of this book—to help you make an intelligent decision on your next optics purchase.

The History of Optics

Generally historians agree that several events or inventions lead up to each other; that is they build, one on the other, and each event or invention creates a need for something else. We know that glass was made in Egypt as early as 3500 B.C. and that crude, though still effective, lenses that are believed to date from around 2000 B.C. have been found in Crete and Asia Minor. Euclid wrote about light and the refraction of light in the third century B.C. In the first century Seneca, a Roman writer, wrote of the glass globe filled with water that could be used as a magnifying glass.

An Arab scientist, Alhazen, conducted a series of experiments in the eleventh century with parabolic mirrors and lenses to magnify objects; his works were translated into Latin in 1572. Still earlier, Roger Bacon (1220–92) wrote about the use of lenses to "read the smallest letters from an incredible distance" and about the power of lenses to make the moon and stars "descend."

Still, it was the invention of the printing press in the fifteenth century, and the demand it created among scholars for spectacle lenses, which led to the widespread use of lenses in Europe. The common belief that the telescope was first invented in 1608 by Hans Lippershey in Middelburg, the United Netherlands, is far from correct. In fact, telescopes were on sale in Paris in 1609 and appeared earlier in various other cities throughout Europe. The telescope was invented by several people at the same time.

Galileo, in Italy, heard of telescopes, perhaps even Lippershey's telescope (of which thirty were made for the military), and he simply reinvented it for his own use, applying basic optical principles. Galileo made his own telescopes, grinding the lenses himself; the largest he made was only 1 3/4 inches across. Using convex and concave lenses in a lead tube he managed to discover the craters and mountains of the moon's surface. He also discovered the moons of Jupiter, the unseen millions of stars in the Milky Way, and the fact that Venus has phases like our own moon. Further, Galileo established that Venus revolves around the sun, contrary to Ptolemaic theory.

All of Galileo's discoveries were made by virtue of two unique characteristics of a telescope: the telescope's magnifying power and its light-gathering power. Galileo upset the entire religious and scientific community of his time and forever changed man's view of himself.

From the seventeenth century on telescopes grew both in size and power. The cost of building these instruments has at times seemed staggering, yet the discoveries and advancements in science combine to make the cost seem a pittance next to the value of expanding man's knowledge. It is important to understand that the instrument Galileo used is not very different from what the average wildlife observer is looking for in a sporting goods store today. The instrument is more refined, it employs the compounded knowledge of centuries of science with the finest manufacturing process we have, yet it relies on two principles of nature: light-gathering power is proportional to the square of the diameter (or aperture) of the lens, and magnifying power depends on the relative focal lengths of the objective glass and the eyepiece glass, which in turn depend on the curvature and composition of those lenses.

I mention this brief history of telescopes, and the small amount of confusion which appears to surround it, because it points out a basic fact: that just a few basic principles apply to all optical equipment. That is, all optical instruments rely on the same principles of physics to function. In the following chapters, as these principles are explained, another, more disturbing fact will emerge—and that is that too often poorly manufac-

This photo of the moon was taken by the author using the Bausch & Lomb 4000 reflecting telescope.

tured optical equipment will be passed off to consumers as fine quality equipment through the use of buzzwords and hype, when in truth the equipment may not be much more effective than the telescope used by Galileo!

As you read through the following chapters and learn about the function of light in optics and how optical equipment today is built to overcome various problems, you will begin to gain an insight into what you are buying. I hope, and it is my intent, to provide you with a reference guide that will help you make intelligent purchases of optical equipment for whatever your need. As a result of your having more knowledge about optics you will find that the dollar you invest in your next binocular or telescope will fetch you the quality you want to do the job required.

2
The Function of Light

I was backpacking on the Rainbow to the Sea Trail in the Santa Cruz Mountains and was determined to make camp near a favorite stream. The trail had been nearly destroyed by a severe winter storm and my usual landmarks were gone. Late in the afternoon I topped a ridge that allowed me to look across several ridges and I thought I could judge the distance between two of the ridges. Guessing at the amount of time it would take me, rather than studying the map, I set off. That was a mistake. I made camp after dark, still several miles from the stream, having underestimated the distance.

Evening light has a habit of fooling everyone. The light reaching a person in the evening is widely scattered by the time it has passed through several hundred miles of the earth's atmosphere, which is, furthermore, laced with millions of tons of dust and other debris. Without the clear shadows of midday it becomes even more difficult to estimate distance than at any other time of day. This problem has given rise to the development of and demand for range finders, which, by using split-image focusing or reticle bracketing, are able to provide a reasonably accurate measure of short-range distance to an object. Some manufacturers are considering marketing long-range devices that will give distances over terrain of several miles. Even the range finder, however, must rely on the same principles of optics as the telescope. All of these principles begin with light.

Light

We are constantly being bombarded with light. I am not referring to the artificial light of the light bulb you are reading by, but to the natural light of our solar system and the stars. Even the stars at night provide us with some light, although it is negligible in its total effect. Yet you see the stars, so there is light. Even on the darkest night of the year there is still *some* light from the sun filtering through to earth's dark side because of the atmosphere's bending and bouncing effect.

Light travels at a speed of about 186,000 miles per second.[1] Compare that to the speed of a bullet traveling at 3,200 feet per second (fps). If you could angle a rifle so that the bullet would travel exactly one mile, or 5,280 feet, with an exact and maintained velocity of 3,200 fps throughout its flight, in the time that bullet is in flight, 1.65 seconds, a single wave of light will have travelled 306,900 miles. If we give the moon a distance from earth of 250,000 miles,[2] that ray of light, which left the spot you were standing at the precise moment the bullet was fired, will have gone 56,900 miles beyond the moon! If we maintain the premise that a single ray of light will travel in a straight line (other theories suggest it will not travel in a straight line, but for our purposes it is sufficient to assume it does), until it either strikes something and is reflected or passes through something and is bent, then we can begin to get a handle on optics.

There are just a few obstacles to understanding the role of light in optics. In the first place we must deal with the idea of light being a "wave." We use the term "ray" in a loose context

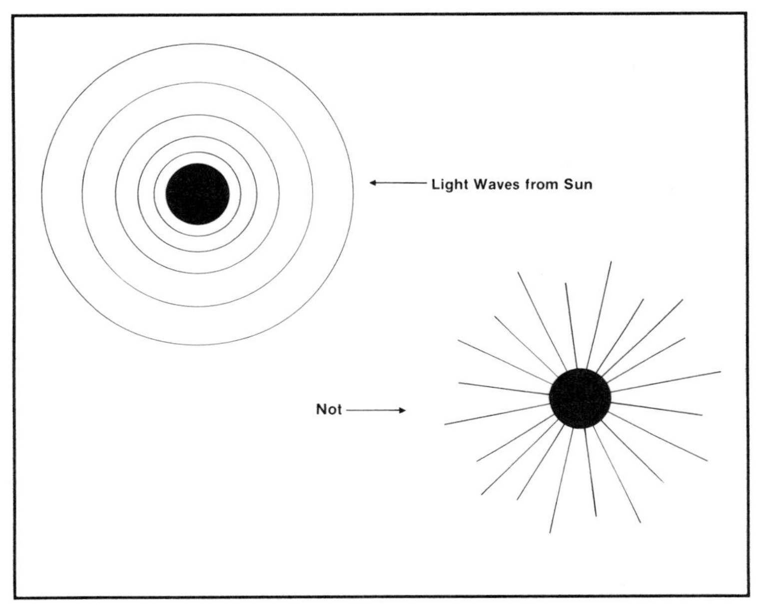

Fig. 2.1. Light is traditionally pictured as rays leaving the sun in a straight line while in fact it is actually light waves spreading outward as circles of light.

because it gives us a better handle for dealing with the effects and desired effects of optical equipment; but for the sake of future discussion, let's understand that light energy is in the form of waves. Also for the sake of our discussion, and the elimination of other complicating issues (in particular, night vision), we will think of all light rays as being produced by the sun.

As the sun burns it produces waves of light. If you think of the sun as the point at which a rock is dropped into the water in a calm pond, and the concentric rings formed by the rock's disturbing the surface as light waves, you have a clear picture of how light leaves the sun: it does not leave as a single ray but in constant waves.

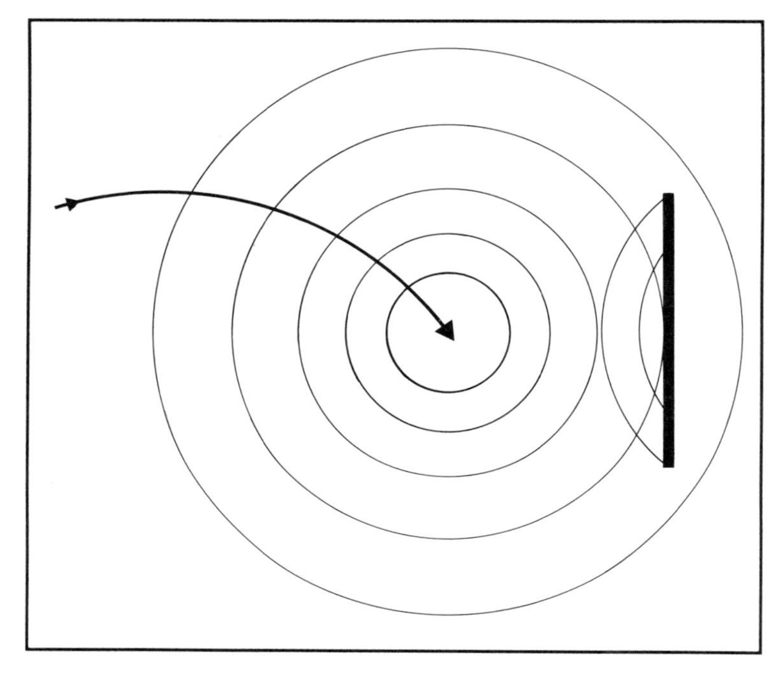

Fig. 2.2. Light waves react in much the same manner as the surface of a pond when a rock is dropped into the water; the waves spread out as concentric circles until they are reflected back.

As these waves spread further and further out from the point where the rock was dropped (the sun) they begin to flatten and spread out. They are also a straight line *in relation to the smallest possible point on the wave.* If you were to examine a one-inch-long section of one of these waves, in a wave that is 10,000 miles in diameter, and your size was such that one inch was equal to ten miles, the wave would appear at that point to be straight. I bring this point up to emphasize that a telescope or other piece of optical equipment is not bending rays of light (as drawn on a piece of paper as straight lines) but bending *waves* of light.

This bending process is the first difficult concept in optical theories. Let's go back to our single ray of light that took off at the same instant the bullet was fired. The bullet is solid and if it strikes an object, there will be two effects—one on the object and one on the bullet. If the bullet is able to penetrate the object it will, but because of the bullet's size, weight, and speed it will destroy the object. If a glass lens is placed in the bullet's path the bullet will pass through it and at the same time destroy the glass. Of course there would be an effect on the bullet as well, and that is, it would slow down considerably and its path would be changed.

Now suppose the light wave strikes an identical lens. It too will pass through the glass but without destroying it, because the light, rather than being a solid object, is in fact a kind of electromagnetic radiation with wavelengths falling into what is referred to as the "visible spectrum." As the light wave strikes the glass of the lens that portion of the wave passing through the glass is slowed down slightly. By altering the thickness of the glass from the edges to the center the effect of "slowdown" can be regulated so that when the wave emerges from the lens some parts of the wave will be emerging earlier than other parts and can then be spread out or focused together.

Using that same ray of light we can also separate it into its "spectrum." The visible spectrum of light consists of the colors red, orange, yellow, green, blue, and violet. These colors are visible and when combined together make up white light or the light we see every day. Depending on the source of light the white light may be altered from one end of the spectrum to the other. All of this, of course, gets into some quite complex theories that, while they would have some effect on optics, for our purposes are superfluous. What is important to understand is that light waves can be bent and that they consist of colors. These colors can be separated by a prism, then merged back

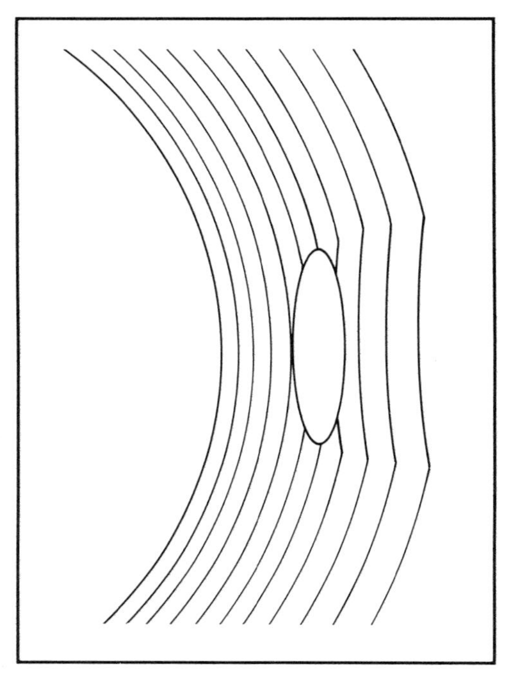

Fig. 2.3. When light waves pass through an object they are slowed down and bent.

together by passing them through another prism. The important fact to remember here is that when light passes through a lens it is bent because one portion of the wave will be traveling slightly faster than another portion after leaving the lens.

Once the concept of light being a wave is understood we can go back and use the term "beam" or "ray of light" for convenience. Consider the ray of light to be that one inch of the ten-thousand-mile circular (concentric) wave and the remainder of the circle is there but not part of our discussion. If you do, the concepts we'll be dealing with will begin to make sense.

There is another aspect of light, and again it is easily understood using the analogy of the bullet. If the bullet strikes

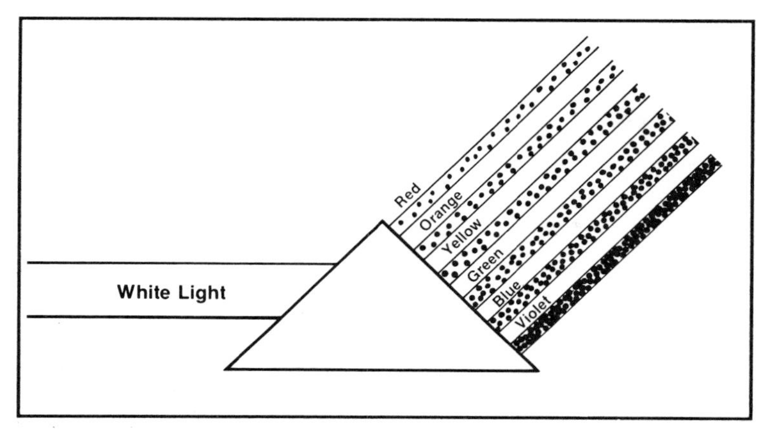

Fig. 2.4. Although light is seen as white it is actually made up of different colors, each of which has its own wavelength. A prism, as shown here, separates the white light into its individual colors.

something it cannot pass through, then it will ricochet off and travel in another direction. Light will do exactly the same thing. Because we know the portion of the wave we are dealing with will react in a prescribed manner when striking an object, we can predict what will happen when certain objects are placed in the path of the wave. If you go back to the pond, and rather than having the entire pond free of obstructions place a single, flat, six-inch board in the water five feet from the point where the rock is to be dropped, then drop the rock, you will see waves emanate outward in concentric circles until part of the wave strikes the board. The six-inch section of wave that struck the board will be instantly reflected back toward the rock forming a second wave. Also, behind the board the wave will immediately rejoin itself, maintaining the integrity of the circle. The reflected waves, because they are not focused, will continue to spread out. Also, because a portion of their energy was used up in the reflection back from the board they will rapidly diminish. These reflected waves are much the same as the reflected light

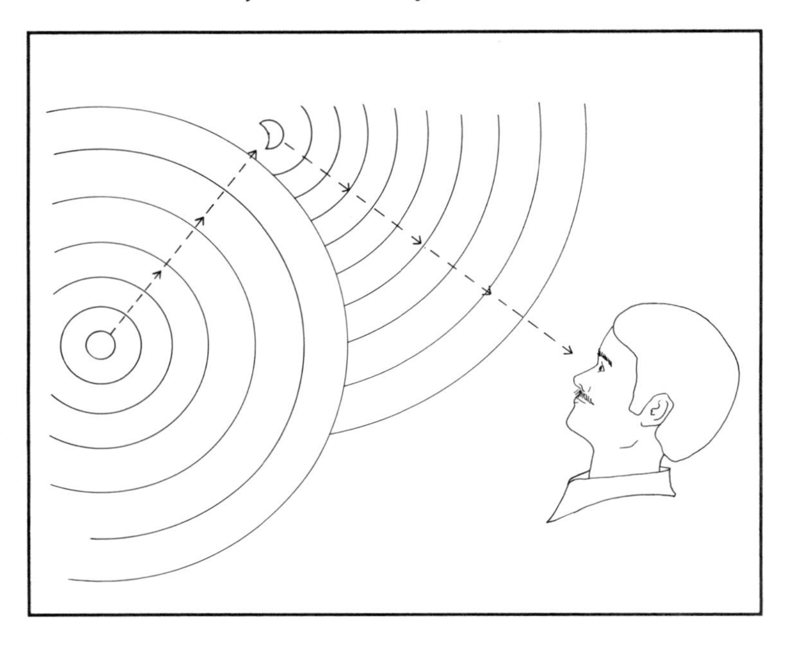

Fig. 2.5. When light strikes an object it cannot penetrate, it is reflected off that object just as it is reflected off the moon's surface.

from the moon. Obviously, some of the energy in the light reaching us at night is lost, absorbed, and scattered. There is little difference between the board and the moon.

Rays of light which are manipulated through glass must eventually end up in one place, and that is your eye. When these light waves reach your eye they don't just stop but are reflected back. In the nanosecond that a light wave is in contact with your eye, the vibrations or wavelengths of the various light waves set up reactions within the cells of your eye's retina. The various wavelengths all serve to produce different chemical reactions within the mechanics of your eye sending chemical signals to the brain. Depending on what cells in your eye are activated and how often in that split-second of time, you will see specific color combinations. Since the light waves are a continuous process, whose component events are actually separated by infinitely small amounts of time, your eye is able to pick out many of the imperfections of various optical equipment.

Light plays a continuous role in optics, beginning with the sun and ending with a chemical reaction in your brain telling you that light waves of various wavelengths are being transmitted to your eye. It was an incredible feat of humankind that light waves traveling millions of miles through space, then bent only slightly by a few pieces of glass, changed the course of history. And yet, when you pick up a telescope, binoculars, or other optical instrument you are doing nothing more complex than trying to get a clear image of a distant animal or scene. It is interesting to note that the desire of Galileo three hundred-plus years ago was not much more profound: he simply wanted to get a clearer view. Light provides that clearer view when passed through lenses and focused for the eye.

Notes

[1] The speed of light is 186,281 mps in a vacuum.

[2] The moon's distance from earth is about 238,851 miles.

3
The Lens

Although you may not realize or want to admit it, a very big part of your outdoor equipment budget will be, or is, directed toward optical equipment. All of these pieces of equipment falling under the category of "optics" have some very common denominators and one of these is *the lens system*. Without lenses your optical equipment is a very expensive hunk of metal and plastic.

The importance of lenses in optical equipment was hammered home for me several years ago. Although I had taken what I considered to be reasonable precautions with the equipment, I wasn't prepared for what happened when I spotted a herd of pronghorn on a distant hill. I could see that they had the spring fawns with them and wanted a better look at the fawns running on the hillside. When I raised my binoculars I could see nothing but black. Somehow, during the rough Jeep ride that afternoon the objective lenses were knocked out of the binoculars. All I could do was watch the hillside with my unaided eyes.

A lens is a piece of glass or other transparent material which has polished, curved surfaces and is used for controlling light. Although, as I pointed out in chapter one, the ancients knew of the magnifying properties of glass or crystal spheres they did not make any deliberate use of lenses until the thirteenth century.

Focusing

One of the first properties of lenses which must be understood is that of *focusing*. We have already established that light is a system of waves and that at a great distance from the source of the wave the light becomes a flat front. As these waves reach a transparent lens an odd thing happens. That small section of the wave striking the lens enters the glass uniformly, slowing down *inside* the lens while the portion of the wave outside the lens continues to travel at its original speed.

In the lens, if the front and back faces are parallel to each other, then when the wave reaches the far end of the lens it will resume the original speed uniformly throughout that portion of the wave that passed through the lens, however it will be slightly behind that portion of the wave that did not enter the lens. We can manipulate the portion of the wave inside the lens if we take that lens and alter it slightly so that it is no longer flat on both sides but is slightly *convex* on the far side, so that the edges of the lens are closer to the first side than to the center, the portion of the light wave reaching the edge of the lens is released a little before that portion of the wave still caught in the center of the lens. As a result, when the center section of the wave emerges from the lens it will be slightly behind the remainder of the wave and will, in effect, be pulling inward on the wave. As the wave continues on it will gradually converge on itself because of this "pull" or curve inward. The point of this convergence is called the *principal focus* of the lens.

The opposite type of lens is the diverging or *concave* lens. In

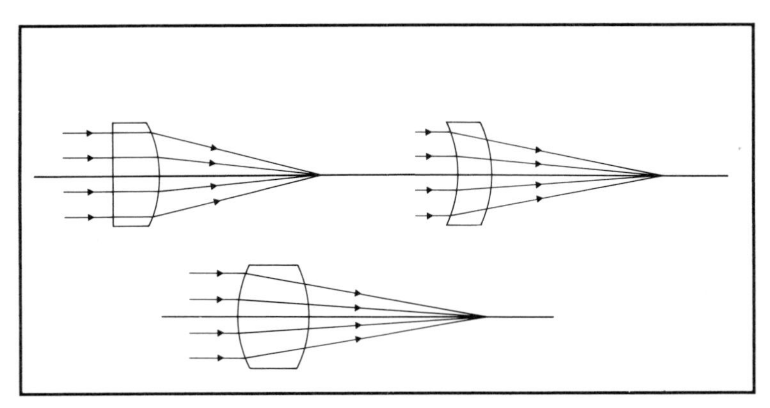

Fig. 3.1. Examples of convex or converging lenses.

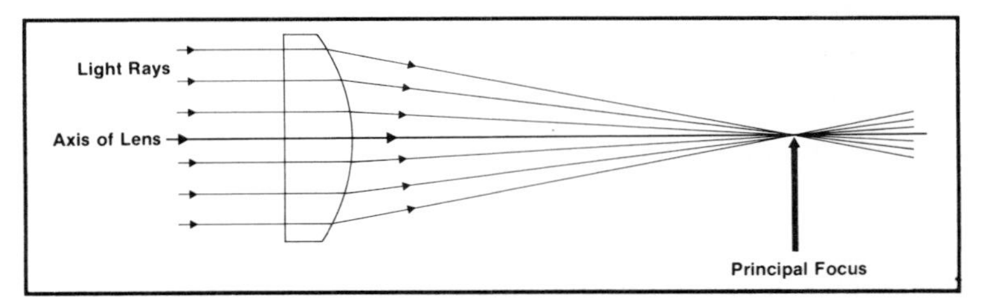

Fig. 3.2. Example of how a lens that is properly focusing all of the rays of light entering it should have only one principal focus (P_f) point.

this case the light wave reaches the center of the lens first and resumes normal speed while the edges of the wave are still passing through the lens. The result is that when the light wave is entirely free of the lens it tends to spread out.

Rays or Waves?

Once we understand these two basic kinds of lenses and the way they affect light waves we can move to another topic. At this point we can switch our discussion from light *waves* to light *rays*. We do this because thinking of light as a ray rather than as

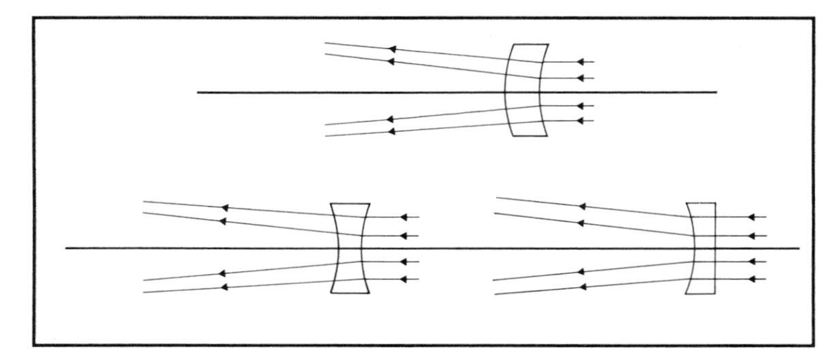

Fig. 3.3. Examples of concave or diverging lenses.

a wave makes it easier to conceptualize the optical principles and their accompanying formulas.

We can describe the behavior of a light wave by drawing several lines perpendicular to the wave fronts. These rays, as they are shown throughout the remainder of this book, do not really exist because the narrowest beam of light you could produce would still have some width and would tend to spread out as it traveled further from the source. A familiar example of this is the common flashlight.

Refraction

Throughout this book we deal with the changing of direction of light rays. Whenever light's direction of travel is changed as it passes through any substance, such as water or the glass of a lens, the effect is known as *refraction* (fig. 3.4). Because light is subject to certain laws of physics, this refraction can be precisely calculated (or predicted) and even the most complex lens system can be "predesigned." These various mathematical formulas and rules enable designers of optical equipment to produce superior equipment at a fraction of the cost of even 20 years ago. What we are concerned with here is not the formulas for

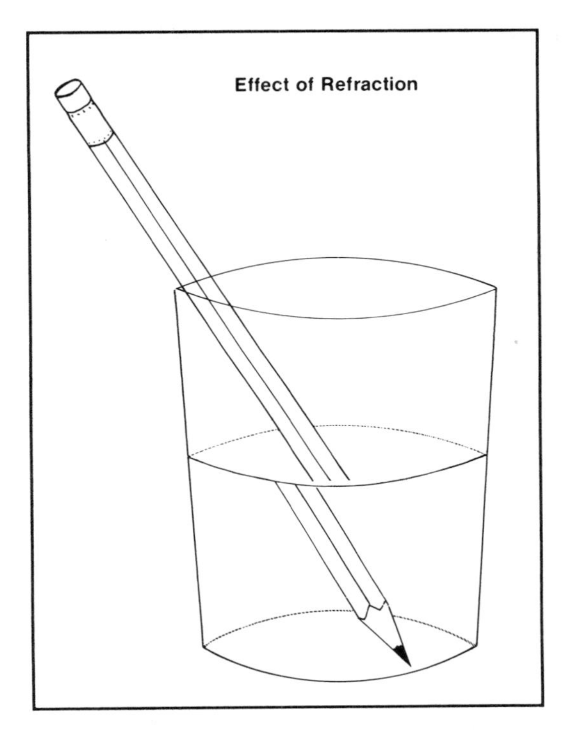

Fig. 3.4. The most common example of refraction is the pencil that appears broken in the glass of water.

designing the various lenses and lens systems but the final product of these formulas. In order to understand this final product, then, you do need some background in the various problems that lens makers must deal with.

The Image

Whenever we buy an optical instrument, whether it is a telescope or binocular, we are buying a piece of equipment that will provide us with an image that is both clearer and larger than that image we see with our naked eye. In optics there are two types of images formed by optical equipment (fig. 3.5). The first

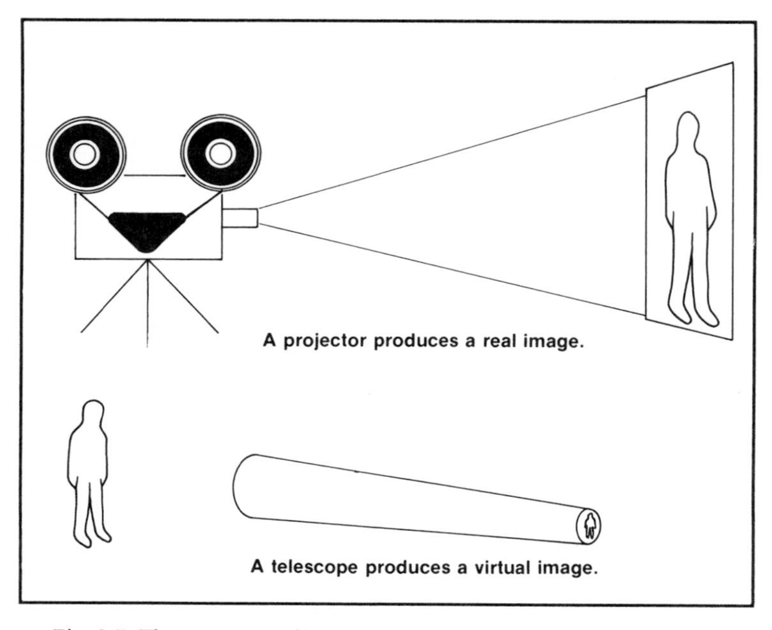

Fig. 3.5. The two types of images produced by optical instruments.

image, called the *real image*, is formed outside the optical system. An example of a real image is the movie projected on the screen in a motion picture theater. Another example of the real image is the image that is received on a piece of film in a camera. The second type of image is the *virtual image*; this is the image seen *inside* the rifle scope, telescope, or other optical instrument. In this case the image is seen *only* by looking into the instrument.

An image is formed by the rays of light leaving an object, being collected by a lens, then focused by that lens at another point beyond the lens. It is the basic property of lenses that all the rays of light leaving an object and passing through a lens, even those at an angle, will be focused to intersect the rays traveling parallel to the axis of the lens at a specified point along the axis. The point of this intersection is the *principal focus* (pf) of the lens.

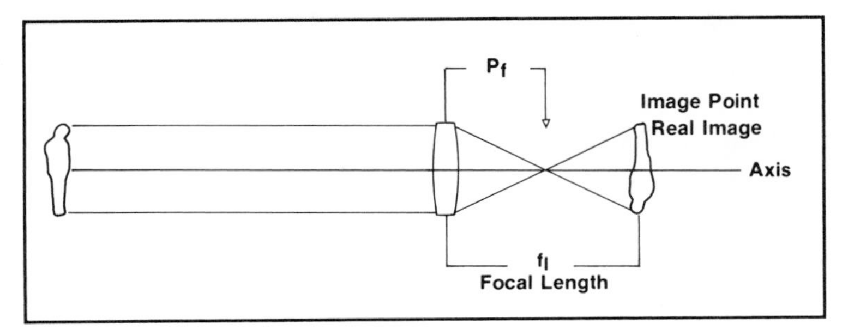

Fig. 3.6. The Principal focus (P_f) of a lens is the point at which the rays of light converge, while the focal length (F_l) is the distance from the lens to the point at which the image is seen and in focus.

These rays continue beyond the principal focus and reform an image of the object at a second intersection that is called the *image point*, popularly known as *focal point*. The *focal length* of the lens then is the distance from the center of the lens to the point at which the image of a distant object is formed. The image that is formed at the focal point of the lens is called a real image.

The real image is one that is formed from light rays coming from an object and passing through a lens to reconverge at the image point and forming a duplicate image of the object. If the distance from the lens to the object is more than twice the focal length from the lens the image formed by the lens will be inverted and smaller than the object being viewed. As the distance between the lens and the object is reduced the image and the object begin to reverse roles. When the object is exactly twice the distance of the focal length of the lens from the lens, the image formed will be the same size as the object. When the lens is moved closer to the object there is an exchange of relationships. Once the object is closer than the focal length a *virtual image* of the object is formed. The rays of light appear to be coming from a point behind the object and the object is no longer inverted but upright and enlarged.

While the real image, the one that is formed when the object is further away, can be projected and focused on a screen or piece of film (although it will be upside down) the virtual image cannot be projected and can only be seen by looking into the lens. You can examine the two types of images and their relationship with a hand-held magnifying glass. Hold the glass up to eye level and look at an object several feet away (such as a shelf of books) and move the glass away from the eye toward the books. You will see the books come into focus, however they will be upside down and smaller than they actually are. Without moving the glass from its position in front of your eyes walk toward the books and you will see them begin to blur in the glass then reappear rightside up and enlarged. The first inverted image you saw was a real image and the second one was a virtual image.

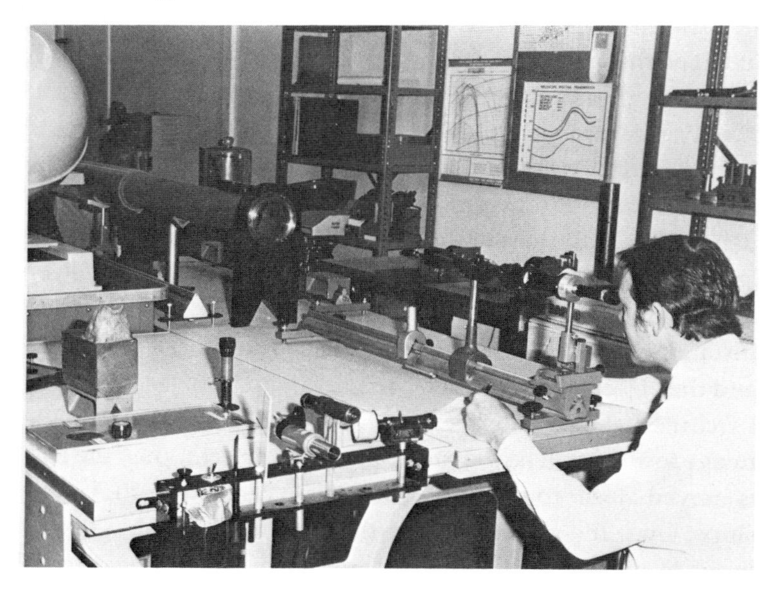

The lenses of modern optics are tested and retested, even in completed instruments. (Photo by Bushnell)

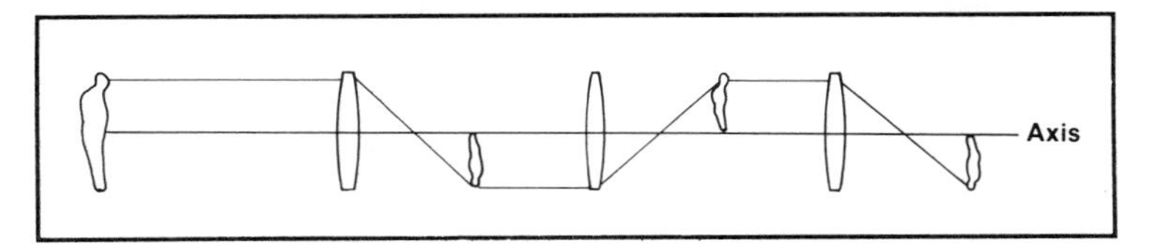

Fig. 3.7. An example of how a lens system works.

The importance of understanding the two types of images formed by lenses becomes apparent when we start working with lens systems. The imaging action of the lenses is determined by treating them in the order they form images. The image formed by the first lens, whether it is a real or virtual image, is treated as the "object" by the second lens, and the image formed by the second lens, regardless of type, is treated as the object of the third lens, and so on through the system until the image reaches its final point, whether it is a real image, such as that which is put on photographic film, or a virtual image, such as that seen in a telescope (fig. 3.7).

Lens Materials

Today's lenses are made from a variety of materials, including plastic, although the highest quality material is made from carefully manufactured types of optical glass. These different optical materials are distinguished by their ability to bend light (the *refractive index*) and to separate light into the different wave lengths (*dispersive power*).

In the early days of lens construction lenses were made from selected pieces of window glass or the glass used to make blown tableware. In the early 1800s the manufacture of a special clear glass for optics was started in Europe. The glass was stirred very slowly while in the molten state to remove as many striations and irregularities as possible. The glass was then cooled slowly and broken up into pieces for lens making.

The broken pieces of lens glass were placed in molds of the approximate size of the lens and then the temperature was slowly raised until the glass again melted, this time into the mold; the lens was then carefully annealed.[1] Over a period of time various chemicals were discovered which could be added to the molten glass to produce a variety of desirable properties. Lead oxide was found to raise both the refractive index and the dispersive power of glass. Barium oxide would raise the refractive index without increasing the dispersive power. Since 1950, lanthanum glass has been used in the production of high-quality photographic lenses.

The cost of optical glass for today's lenses varies considerably. In the low-cost equipment marketed by some firms plastic is used instead. Plastic lenses are used exclusively in some simple cameras. One of the most popular plastics is polymethyl methacrylate. Whether a lens is made from plastic or glass it is still subject to the many variations and "faults" found within optical systems.

Mirrors and Prisms

Before moving on to the next chapter and covering the problems or "faults" of optics, which must be overcome to produce an instrument, we'll take a brief look at some of the other components of optical systems. One of the most common is a mirror.

Curved mirrors form images using geometrical principles similar to those which apply to lenses. The major difference is that the image will be on the same side of the mirror as the object. The principles are derived from the law of *reflection*. The incoming and reflected rays of light will make equal angles with the radius of the mirror surface at the point of reflections and both rays will be on a plane that will include the radius. The focal length of a spherical mirror is half the radius of the curvature of

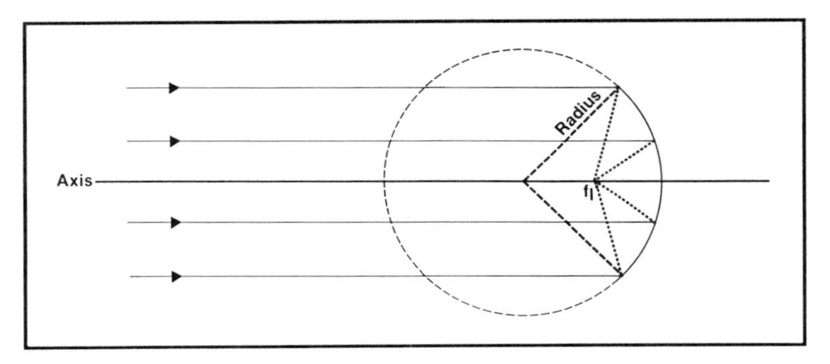

Fig. 3.8. Example of how a reflecting mirror works. The focal length (f_l) of the mirror is one-half of the radius. The angle of the reflection will make an angle equal with the radius.

the mirror. Rays parallel to the axis are reflected through the principal focus and the rays through the center of the curvature of the mirror are reflected back on themselves (figs. 3.8, 3.9).

The use of mirrors becomes very important when building optical instruments for which weight and bulk are prime considerations. Because the reflective laws can be used to gather more light (see chapter five), and then to focus that light where it is picked up by a lens or lens system, a more powerful optical instrument can be built with both a weight and a cost savings over a "traditional" system.

Prisms are also used in optical systems to reflect and bend light along specific, predetermined angles for the purpose of either transmitting the light rays to a lens in a more compact instrument or increasing the focal length in a compact unit. There are a variety of prisms in use, however the most popular is the Porro prism.

Both mirrors and prisms can be used to reflect and bend light waves along various routes through an optical instrument for a variety of purposes. As we cover the various types of optical instruments the importance of quality in both the mirror

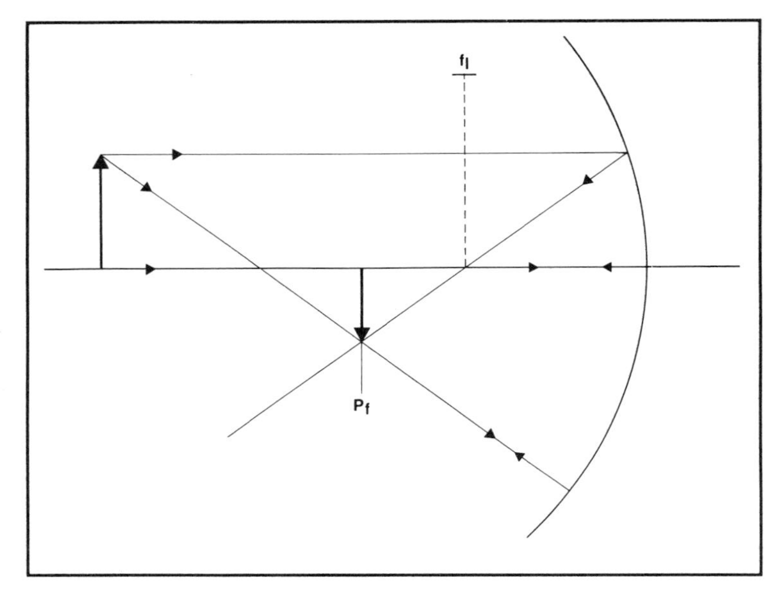

Fig. 3.9. Illustration of a reflecting mirror showing the principal focus (P_f) and the focal length (F_l).

and the prism, and the problems which can be expected, will become more important.

Optical systems are by their nature complex instruments. Even the most inexpensive unit will be somewhat complex. If there has been any effort at all to solve the problems we will cover in the next chapter, then the price of the unit will go up. You cannot solve optical problems cheaply. The kind of quality demanded by even casual users of optical equipment leads to more expensive equipment. It is interesting to note, however, that once the uses and potential problems in optics are understood it is possible to buy a perfectly functional piece of equipment that will meet your needs without spending this month's house payment. To find it you need to know what to look for.

Note

[1] Annealed: a process of cooling slowly under controlled conditions.

4

Optical Faults and Myths

One piece of advice I like to give people who ask me to help them select optical equipment is to tell them to take their potential purchase outside and check it under the light of the sun. Indoor light is not white but is saturated with other colors which will make the light you see through the equipment appear slightly greenish to red. An example of this is easily found by simply taking a color photograph indoors using one of the outdoor films, such as Kodak's Kodachrome 25 slide film. Outside this film will produce beautiful photographs under a bright or hazy sun. Bring the same film inside and you must use either a filter over the lens to correct the color of the light passing through the lens and onto the film or a color-corrected flash.

In earlier chapters I explained that light is made up of waves and that different colors of light consist of waves of different wavelengths. Taken together, however, these waves produce white light. The prism with two 45 degree sides can split this white light into the various colors: red, orange, yellow, green, blue, and violet. I have also described how a lens manipulates light waves (rays). It would appear, then, that since a lens is an optical instrument that focuses and bends light, which is furthermore composed of waves of different lengths (colors), there is an excellent chance that something is taking place which could create problems for both the designer and the user.

There are actually six problem areas or "faults" in any lens which must be dealt with by optical equipment manufacturers. Through the combination of different glass types, designs, and distances separating the various elements or lenses, these faults can be dealt with and corrected to a degree. The amount of effort put into correcting these faults is what separates the junk from the quality optical equipment. The more engineering put into solving these problems the better the equipment and, of course, the higher the price tag. If you know what the problem areas are and have some idea of how they are dealt with in a lens system, then it is possible to match the amount of money you plan to spend to the equipment being purchased. By knowing which faults will cause you problems under a given set of circumstances you can make an informed selection.

The six "faults" in optics are:

1. chromatic aberration,
2. spherical aberration,
3. coma,
4. distortion,
5. curvature of field, and
6. astigmatism.

As each fault is described the problems produced by that fault will be explained.

Chromatic Aberration

As you recall from the discussion of the prism and white light, a ray of light is easily separated into its various colors (or wavelengths). Because some of the wavelengths (red) are easily

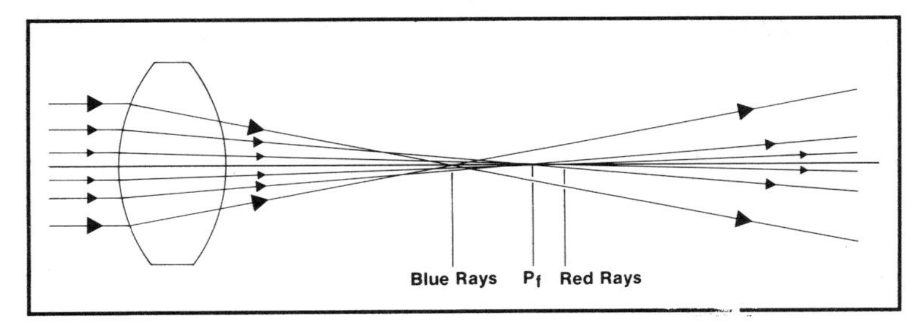

Fig. 4.1. Chromatic aberration is caused when a lens is unable to focus the various colors of the spectrum on the principal focus of the axis of a lens.

separated, requiring very little of the angled glass, there is no single lens which is capable of focusing all the colors at the same point. While the principal focus point and the image focus point will be fairly close, on closer examination you will see, with any single lens, a slight color "halo" around the object. This halo is formed because the lens cannot focus all the rays at the same point (fig. 4.1). Some, like the red rays, are focused in front of that point while the blue are focused behind it.

To correct for this *chromatic aberration* two lenses are combined; one is a strong lens of a low-dispersion glass and the other is a weaker lens made of high-dispersion glass. A combination of two such lenses is then said to be *achromatic*. This sophisticated-sounding word is exploited in the manufacture of inexpensive equipment. Although the first lens of an optical instrument may be achromatic this does not in itself solve the problem of chromatic aberration. Remember that the second lens of a system (such as a refracting telescope) will pick up the image formed by the first lens and treat it as the object and then form its own image. In order for the second lens to provide an image that does not itself have chromatic aberration it too must be achromatic or the instrument will still suffer from this fault. Still, one of the most popular buzzwords used in sales hype to lure customers to buy inexpensive lenses is "achromatic lens" or "lens system." In order to spot this fault you need to take the instrument into the sun where it is receiving the full benefit of white light and look at various distant objects that present a clear view of their outside edge. If there is a halo effect around the object, then the instrument is not fully corrected and you may be paying for something that does not exist.

There are five more faults whose seriousness in a lens can render it useless. The five remaining faults are listed under the grouping of *Seidel sums.* If a lens were perfect and the object it focused on was a single point of monochromatic light, the light emerging from the lens would be a portion of a sphere and would be centered about the image point. This simply does not occur with lenses. The failure of the light wave to form the image at the ideal image point is called the *OPD* or the *optical path difference.* Within this difference are the five remaining faults.

Spherical Aberration

This fault refers to the inability of a lens to focus all the light rays from an image on the same place of the axis of the lens. This is caused because each small section of a lens will want to focus the light rays passing through it at a slightly different point. The problem can be eliminated by careful manipulation of the focal plane of the lens by applying a longitudinal shift to the focal plane and adjustments in the lens surface itself to control the focus point. If the problem is not corrected within a lens the result is you cannot obtain a sharp image in the instrument. Obviously, this is one of the master headaches for lens designers because this problem must be solved at the same time the designer is dealing with chromatic aberration.

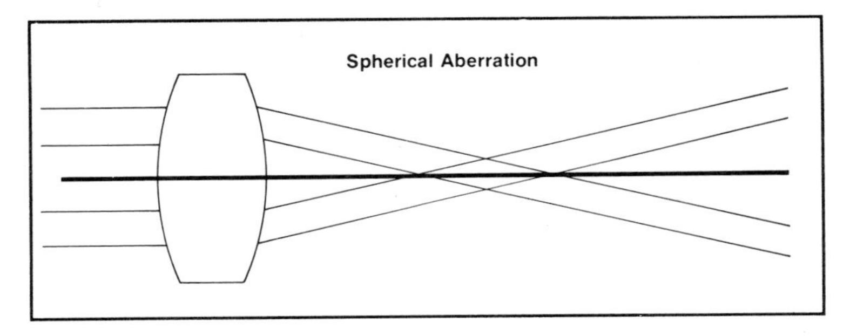

Fig. 4.2. Spherical aberration occurs when a lens does not allow the rays of light entering at different points along the lens to focus at the same point on the axis of the lens.

Coma

Coma is another form of spherical aberration. It is slightly different in that it is produced when light enters a lens diagonally. Each circular zone of a lens tends to form a ringlike image in the focal plane, resulting in a series of successive concentric zones. The effect of these zones in viewing is to cause a cometlike appearance in the outer parts of the viewing field of the instrument (fig. 4.3).

When the two problems of spherical aberration and coma are solved in a lens the lens is then called *aplanatic*. In any quality instrument the lenses must be both achromatic and aplanatic. Both of these faults are easily checked for in an optical instrument before it is purchased.

Another buzzword that is often used in promoting lenses is *orthoscopic*. This refers to a lens being free of distortion (another aspect of the *Seidel sums*). When distortion is present in a lens the entire image point is displaced toward or away from the axis of the lens. This is caused because the magnification of the lens is

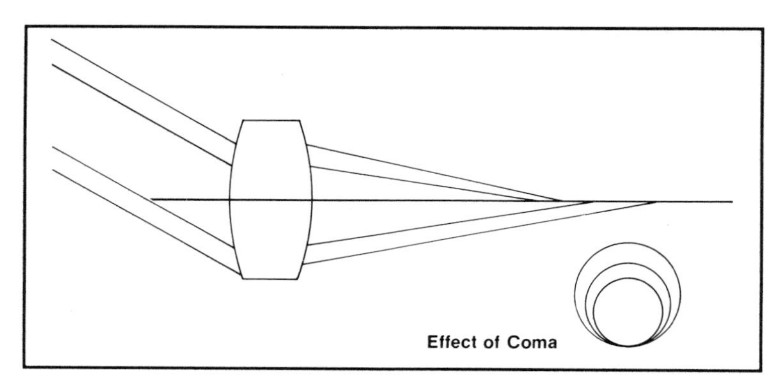

Fig. 4.3. The effect of coma occurs when a lens is not corrected to allow rays of light entering a lens at an angle to intersect the axis of the lens together.

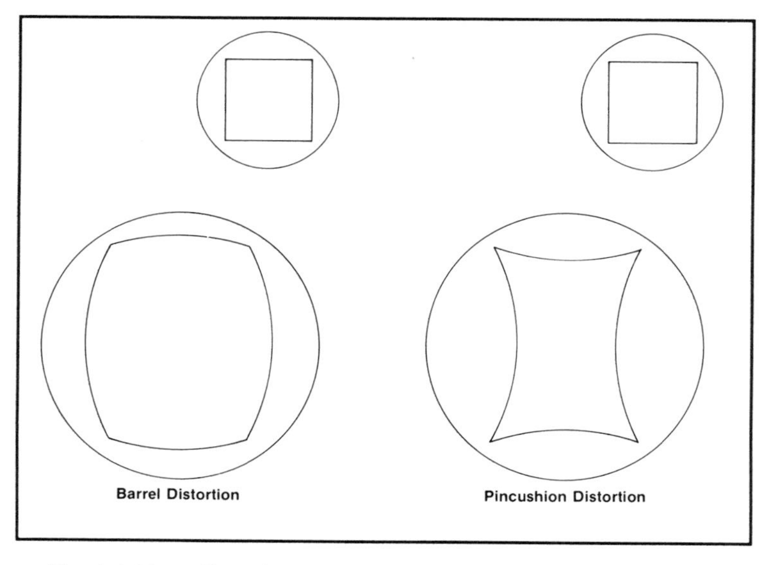

Fig. 4.4. Two effects that are frequently found in low-quality optical instruments are barrel distortion and pincushion distortion.

not uniform over all areas of the lens. A square viewed through a lens which is not corrected for distortion would appear to be slightly rounded or barrel shaped (fig. 4.4). It is barrel shaped because the lens's magnification decreases toward the edges of the lens. If the square is squeezed along its edges making the corners form points, this means that the magnification of the lens increases toward the edge; the effect is called *pin-cushion distortion*. The amount of distortion in a lens can vary greatly, from a slight amount in a medium-priced instrument to a real problem in very inexpensive scopes.

Curvature of Field

The problem of *curvature of field* is operating when it is possible to focus the center of an image but not the edges (fig. 4.5). This fault is somewhat common in lens systems where one

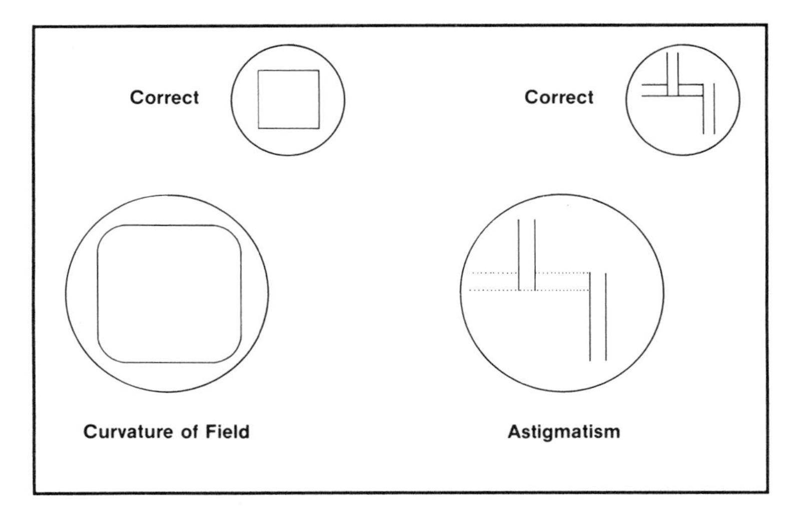

Fig. 4.5. Curvature of field and astigmatism. The curvature of field shows rounded edges where square edges should be seen in the instrument, and astigmatism is apparent when lines that are at right angles to each other cannot all be brought into focus at the same time.

or two of the internal lenses are not corrected to the other lenses. The curvature of field is solved by combining two different types of glass, thereby combining negative and positive elements (low and high index glass) to reduce the curvature.

Astigmatism

Astigmatism is found in the human eye and also is a common optical defect; however it is usually so slight that it goes undetected. What this refers to is the inability of a lens to focus on two lines that are at different angles (fig. 4.5). When a lens has been corrected for this it is called an *anastigmatic* lens.

Of these six faults the most common are chromatic aberration and spherical aberration; they affect the center of the image seen in the instrument. The other faults appear as a gradual loss of detail: the focus from the center of the field to the outer edge will gradually deteriorate. By knowing what to look for and not falling for some sales hype ("that color you see around the thing you're looking at is the special coating on the lens . . ."), you can obtain your dollar's worth.

Everyone's Alibi: Parallax

The "hunters alibi"—whereby one explains away a poor shot—really refers to the distortion effect known as *parallax*. This same effect is sometimes cited by the amateur photographer to explain why someone's head was "chopped off" in a photograph. Whenever someone needs a handy and official-sounding excuse for something that went wrong involving optical equipment, they like to fault the instrument's parallax. All parallax really is is the *apparent* movement of the object in a scope's field of view relative to the center of the viewing area. You can see an example of this by holding your hand in front of your face and forming a circle with your thumb and finger. Look through the circle at an object, then, without moving your hand, move your head from side to side. The object appears to move when in fact it does not, nor does your hand. In an optical instrument parallax is occurring when the primary image of the object seen through the scope is focused either in front of or behind the reticle, or when the eye is off the optical axis of the scope. If the primary image is on the same plane as the reticle (crosshairs) of a rifle scope or the focusing glass of a camera which is the surface where the image is brought into sharp focus before taking the picture, or if the eye *is* on the optical axis of the scope, regardless of where the primary image is, there is no parallax.

In low-quality, inexpensive scopes, however, parallax can be serious and actually is a good excuse for missing a shot. If a lens is not corrected for spherical aberration or astigmatisim, the image may form at some point within the scope away from the reticle creating a serious parallax problem. When a reticle or focusing glass is loose in an instrument and can move even a fraction of an inch parallax can be created.

It is important to realize that parallax cannot be completely eliminated in scopes, so manufacturers design their shooting scopes to be "parallax free at a set distance." For most rifle scopes this is 100 yards. There will be some parallax at any range closer or further than 100 yards. You can easily check the parallax in a scope by holding it steady while sighting on an object, then moving your head from side to side or up and down so that the object remains within the field of view. The apparent movement of the object is parallax and the closer or further from 100 yards (or whatever the scope is set to be parallax free at) the object, the greater the parallax will be.

If a shooter takes the time to line up on the scope by taking the proper sighting through the scope, parallax will have very little, if any, effect. The same is true for the photographer using a telephoto lens. The eye should be aligned with the axis of the lens. Rather than being an excuse for a failure, the claim of "parallax" is usually a statement about poor habits.

Once these problems or faults are understood for what they are and how they affect all optical equipment, we can begin to study the real issue—how we see what we see through the lens. I am sure all of these official-sounding names and titles are confusing, but they point to one fact: when buying any optical instrument look for the *clearest possible image* and don't be led down a path by a sales clerk who is bent on making a sale.

In chapter five we will look at what "corrected" lenses and rays of light are actually doing in the optical instrument.

5
Understanding the Equipment

As part of the research for this book I was provided with a selection of binoculars by Bushnell/Bausch & Lomb for testing in a series of hunting camps by hunters and in retail stores by general public. What I was looking for was information about the "why" of the selection of optical equipment.

My test binocular selection arrived the day before one of the hunting camps opened. I packed the binoculars in the trailer and headed for camp. The night before the season opened I spread the selection in front of the eager hunters who had agreed to help with the test and let them select the pair they wanted to carry while hunting the next day. The only requirement I made was that each man let me know why they selected the binoculars they took and what their feelings were about them at the end of the hunting day. The next day they were to select another pair and repeat the test.

When the hunting camp was over I took the binoculars to a sporting goods store to leave the entire selection where it could be looked over by the customers and commented on. Again we asked for consumer input on selecting binoculars, only this time we asked customers to select two pairs without considering price—one for general recreation viewing and the other for observing wildlife only.

Although the response sample I obtained represents a very small segment of the population, the respondents were consistent with their selections and comments; thus we can draw some conclusions about how binoculars are selected and can compare these responses to what should be some actual considerations when buying optical equipment. The overwhelming criterion, and one that seems to make more of an impression on the consumer than any other, is appearance. If the buyer has the money to spend, he seems to be more concerned with the appearance of the equipment than how well it would perform outdoors, expecting performance to follow appearance. Here are the two strongest arguments for buying.

1. Most outdoor enthusiasts based their selection on appearance and if the glasses looked rugged. Rubber armoring was more important than other considerations.
2. Most general purpose (recreation viewing) selections were based on power.

Neither reason is, in itself, justification for the purchase of any optical instrument. To understand why, you need to know what the various numbers printed on optical instruments are and how they translate into performance.

Objective Lens

First, light is collected by the big lens at the front of the piece which is called the *objective lens*. This lens's diameter is measured in millimeters. Some common diameters for objective lenses are 20, 24, 35, and 50 mm. There are lenses which are larger than 50 mm and smaller than 20mm. The larger the number the bigger the objective lens is. In an ideal situation, that is, where all other factors on different instruments are the same, the larger the lens the more light it can collect. Let's look at the possibilities for four common objective lens sizes: 50 mm, 35 mm, 24 mm and 20 mm. To find out how much more (or less) light is gathered by each lens *at the start* of the viewing process we'll use the formula for finding the area of a circle: πr^2 (π=3.14). Since the radius (r) is one-half of the diameter (D), the radii of the four common objective lenses are 25 mm, 17.5 mm, 12 mm, and 10 mm, respectively.

Once we know the radius of each lens we can apply the formula to find the *surface area* of the lens.

50 mm Objective Lens: r^2=625 $\pi \times 625$=1963.5

35 mm Objective Lens: r^2=306.25 $\pi \times 306.25$=962.11

24 mm Objective Lens: r^2=144 $\pi \times 144$=452.39

20 mm Objective Lens: r^2=100 $\pi \times 100$=314.16

Knowing the surface area (A) of the objective lens does not, in itself, provide useful information except in giving you some idea as to why a larger lens collects more light than another, smaller lens. To really compare the lenses we'll take the above information two steps further—first by obtaining the number of times one lens is larger than another, and second by building a comparison chart of the lenses. To obtain the comparison between the lenses divide the surface area of the larger lens by that of the smaller lens. I've divided out the standard lenses in the following chart.

D=50/A=1963.5

$\frac{1963.5}{962.11}$ =2.04 *times* larger (35 mm lens)

$\frac{1963.5}{452.34}$ =4.34 *times* larger (24 mm lens)

$\frac{1963.5}{314.16}$ =6.25 *times* larger (20 mm lens)

D=35/A=962.11

$\frac{962.11}{452.39}$ = 2.13 *times* larger (24 mm lens)

$\frac{962.11}{314.16}$ = 3.06 *times* larger (20 mm lens)

D=24/A=423.99

$\frac{452.39}{314.16}$ = 1.43 *times* larger (20 mm lens)

The 20 mm lens is not listed because it is the smallest of the lenses we are working with. Now that the comparative areas of the lenses are known we can put together a table showing these comparisons.

	20 mm	24 mm	35 mm	50 mm
50 mm	+6.25	+4.34	+2.04	—
35 mm	+3.06	+2.13	—	-2.04
24 mm	+1.43	—	-2.13	-4.34
20 mm	—	-1.43	-3.06	-6.25

Reading the table from left to right you can see that the 50 mm lens has 6.25 times the area of the 20 mm lens and that in theory it will offer that much more light-gathering capability.

The plus (+) and minus (-) signs help to indicate whether the lens on the left has a greater or lesser light-gathering capability than the one listed across the top of the table.

If all things were equal we could say that the 50 mm lens does in fact "collect 2.04 times more light than the 35 mm lens." We cannot make that statement, however, because all lenses are different. Also, light is absorbed by the lens during its passage through the lens. Remembering that light is energy and as energy it will be absorbed to varying degrees by different materials as it passes through or is reflected off the surface, we can see that comparing lenses is a little more complex than we first thought. If we take the amount of light entering the lens as 100 percent of the total light available, we can safely argue that because of absorption, only about 75 percent of it will end up striking the eye of the user. Even more disappointing is the fact that if we took that same 100 percent of light and said it represented the total amount of light *striking* the lens, we could reduce the amount of light *entering* the instrument by 5 percent, at minimum. Once inside the instrument other effects including absorption and *scatter* will reduce the effective amount of light down another 25 percent so that 30 percent of the total amount of light which struck our objective lens *never reached the eye of the user!*

To compensate for this loss manufacturers use coated lenses. In most cases this coating refers to *reflection reducing* coatings that are designed to reduce the amount of light lost both inside the instrument with internal elements and outside with the objective lens. The coating most often used is magnesium fluoride which, when properly applied to the objective lens, can reduce light loss by reflection outside the instrument from 5 percent to 1 percent.

With this loss of the light being collected and transmitted to the other lenses within the instrument the need for the larger objective lens on your field optics becomes apparent. The key point to remember is that the larger the objective lens the brighter the image you actually see in the instrument.

From the Objective to the Eye

After traveling through the lens and prism system light emerges from the instrument and is seen by the eye. Depending on the power of the instrument the object appears larger, or closer and is magnified.[1] Because it is appearing closer (say seven times closer with seven power) it will also be clearer than seen with the naked eye.

When you look through a binocular or telescope the "area" that is seen through the instrument is the *field of view*. Because the amount of area seen is important to the use of the instrument the field of view must be known and understood. The standard of measurement for the field of view can be expressed in either yards and feet or meters. In both cases the standard is the width of the area seen through the instrument at either 1000 yards or 1000 meters. If 1000 yards is used the width will be expressed in feet and for a standard 6× instrument the field of view would be 446 feet. With the same instrument at 1000 meters the field of view would be 149 meters. Instruments with wider fields of view would have wide angle viewing.

Some instruments express field of view in degrees, which is the *angular field of view* rather than yards or meters. The angular field of view is the angle from the center of the objective lens to the edge of the viewing area. To convert this into the field of view in feet multiply the degrees of the angular field of view by 52.5, which is the number of feet in a degree at 1000 yards (degrees × 52.5 = field of view). To find the field of view on a standard 6× instrument with an angular field of view of 8.5 degrees the result would be: 8.5 × 52.5 = 446.25 feet.

With the field of view understood we can move on to

another useful number and that is *apparent field of view*. The area of the image that is seen in the eyepiece when looking through the instrument is the apparent field of view and is expressed in degrees. To find the apparent field of view of an instrument multiply the power of the instrument by the angular field of view. Using the same instrument as above as an example we would multiply it as: 6 × 8.5 = 51 degrees. Therefore the *apparent field of view* would be 51 degrees. A wide-angle binocular should have an apparent field of view of at least 65 degrees. A rifle shooting scope with a wide-angle view should have an apparent field of view of 26 degrees.

Other numbers or terms describing the performance of optical equipment are *exit pupil* (EP), *relative brightness* (RB), and *relative light efficiency* (RLE). Two other terms which need to be explained are *twilight factor* (TF), and *dark vision index* (DVI).

Without expensive testing equipment we must rely on a basic understanding of what is supposed to be happening and what we can see for ourselves in the instrument when we buy optical equipment. The first number we will deal with is *exit pupil*. The human eye has an iris which works much like the aperture on a camera. For average light the human iris is open 2.5 mm and in poor light is open to about 5 mm. The exit pupil on an optical instrument is the diameter of the light rays emanating from the eyepiece. To find the size of the pupil of an instrument divide the objective lens size by the power of the instrument. On a 7×35 binoculars the exit pupil would be 5 which is the result of dividing 35 by 7. Since the objective lens size is expressed in millimeters the exit pupil is also expressed in millimeters.

The average person can use an exit pupil of 2.5 mm to 3 mm in full daylight and 6 mm to 7 mm at night. A larger exit pupil size than what the eye requires or can use allows the instrument to be used in a wider variety of lighting conditions and does not require as exact a placement of the eye in relation to the lens. Larger objective lenses play an important role here because they provide larger exit pupils with more light, which in turn produces sharper image detail for viewing under varying lighting conditions found in the field.

An additional comparison between scope types can be found by comparing a rifle scope to a telescope. A rifle scope equipped with (for example) a 35 mm objective lens and having a power of three would have an exit pupil of 11.67 mm. Obviously you would obtain a bright image through this instrument. On the other hand, one of the popular "compact" scopes found on the market for many .22 rimfire rifles has an objective lens diameter of around 20 mm; if the power of that scope is three, the exit pupil would be 6.67 mm (20÷3=6.67).

Now, compare those numbers to telescopes, which have greater magnification than either binoculars or rifle scopes, and you will find that there is an even smaller exit pupil, which will require more exact placement of the eye to the eyepiece for proper viewing of the image. If a telescope has an objective size of 60 mm and a power of 20, the exit pupil is down to 3 mm. On some variable power telescopes the power ranges as high as 45 and the exit pupil is then reduced to 1.33 mm.

Relative brightness is another term that is often used with little consideration to what it actually means. To determine the relative brightness of an instrument the size of the exit pupil is squared. A 7×35 binoculars with an exit pupil of 5 mm then has a relative brightness of 5^2 or 25. In instruments which do not have any, or very little, coating over their various elements, this number is fine and gives a fair idea of the relative brightness of the instrument. However coatings, if they have been properly applied, will significantly improve the instrument's performance. So we need to have another index that enables us to gauge the instrument's performance. To obtain the *relative light effi-*

ciency (RLE) of an instrument multiply the relative brightness by .5 and add the result to the RB. On a 7×35 binoculars with an exit pupil of 5 mm and an RB of 25 (5^2) you would add 12.5 (.5 × 25) for a total of 37.5; so the RLE of the instrument *with properly coated lenses* is 37.5.

None of these terms really explain one of the key features of binoculars and other optical instruments—their ability to enable a user to "see" detail in darker areas, such as a deer in a shadowed forest. To find some sort of number to define this, the *twilight factor* (TF) was developed and is used extensively in Europe. Instruments such as the Swarovski Optik SL 7×50 Marine Binoculars will include the twilight factor in their literature. To determine TF the following formula is used: twilight factor = $\sqrt{\text{magnification} \times \text{objective lens size (mm)}}$, where √ is the square root. So the TF is the square root of the result of multiplying the power of the instrument times the objective lens size. For the 7×35 binoculars this would be $\sqrt{7 \times 35} = \sqrt{245} = 15.65$.

Although twilight factor is accepted as a useful and informative index of an instrument's performance in the field, Bushnell/Bausch & Lomb has raised serious doubts about it being a definitive factor and their engineers have gone on to create another index which provides the buyer with some interesting data; this is called the *dark vision index* and it is a guide for evaluating the performance of a binocular under poor lighting conditions. The formula for this index is

$$\text{dark vision index} = \frac{\sqrt{\text{objective lens } D^5 \div \text{Power}^3}}{1000}$$

To translate this formula, DVI equals the square root of the result of dividing the fifth power of the objective lens diameter in millimeters by the cube of the magnification (power) and dividing this product by the number 1000.

A 7×35 binocular would be plugged into the formula as follows:

$$DVI = \frac{\sqrt{35^5 \div 7^3}}{1000} \qquad \frac{\sqrt{52,521,875 \div 343}}{1000} \qquad \frac{\sqrt{153125}}{1000} = .39$$

The dark vision index number can be used to rank instruments in the following manner. The closer the DVI is to one the better the system will perform under adverse lighting conditions. Binoculars with a DVI of one or very close to one are excellent for general night or twilight use. The closer the DVI gets to 1.5, the better the optics will be yet the bulk will also increase to the point where the optics will be hard to handle. When the DVI nears two the instrument has gone past the point of diminishing returns and is no more effective than equipment with smaller DVI's because the image is so bright the human eye cannot make use of it. The following table presents the various indexes I've covered here for the four common objective lens sizes (50 mm, 35 mm, 24 mm, and 20 mm) coupled with an eyepiece system which gives a power of seven. Assuming that all four instruments have coated lenses and are equal in manufacturing care and expertise, they chart out as follows.

	7×50	7×35	7×24	7×20
EP[1]	7.14	5	3.43	2.86
RB	50.98	25	11.76	8.18
RLE	76.47	37.50	17.64	12.27
TF	18.71	15.65	12.96	11.83
DVI	.95	.39	.15	.10

[1]Exit pupil in mm.

Although the math may at first seem intimidating, the reason for presenting it is to give you some idea of what any optical instrument should do. If you know, for example, that the DVI of an instrument is .95, and in looking at it you find it seems dim for an instrument of that DVI, then something is wrong. The same is true for the exit pupil. If you know that a 7×35 instrument should have an exit pupil of 5 mm and is being advertised as having "Fully Coated Lenses," yet the image seems dim or irregular, especially in a store, then there is something wrong. By knowing what an instrument is supposed to do and what the ratings being used should tell you, you can make an intelligent selection of the product you need.

Note

[1] Power is covered in chapter six.

6
Power

For reasons that have always evaded me, most people who use optical equipment, whether they are sportsmen, wildlife observers, police officers, or amateur astronomers, tend to rely on power as their salvation. There seems to be a magic associated with power in optical equipment that leads many people to believe that by having more power in their instrument they will solve whatever problems they are having, or that the power will extend and improve their shooting scores, their powers of observation, or make their jobs easier. The truth is there is not a piece of optical equipment made that will solve your problems if the problems are the result of poor habits on your part.

The sportsman who believes a nine power rifle scope will solve his shooting problems, or the wildlife observer who believes his enjoyment of watching wild animals will be enhanced by buying a sixty power eyepiece for a spotting scope, are in for a rude awakening. If the difficulties are a product of poor equipment use techniques, or a physical problem a person is not aware of, power is not going to improve anything or enhance the enjoyment of the outdoors. This is true for *anyone* who uses optical equipment—amateur or professional.

Power and Magnification

The first thing to understand is just exactly what *power* is. Power is a reference to magnification and it can be expressed in optics as a whole number such as 2, 4, 6, etc., or it can be expressed as a decimal such as 1.5, 3.5, etc. These numbers are then followed by an "×," which indicates that the number is the instrument's "power" or "magnification." The "×" is simply a mathematical symbol used to express multiplication such as 2×4=8 or 4×4=16. The "×" is used to replace the word "times," as in "4 times 4 = 16." In optics the number followed by the "×" translates into: "object size is multiplied by (number) times." As an example, an object that is one foot high when viewed without the aid of an optical instrument would "appear" to be four feet high when viewed through an instrument with a "power of four."

This illusion is possible because the entire area viewed in the instrument is also enlarged or magnified and it is only a small part of the area as seen by the unaided eye. Hold a telescope up so you can look through it with one eye and the other is seeing the surrounding area outside the scope. The area seen in the telescope will appear larger than the area seen with the unaided eye. Another important point to remember is that the area seen within the scope is also referred to as the "field of view." To get an idea of how this power works and its effect, let's take an object that is four feet high (about the height of a mule deer) and move it out to 100 yards. For the purposes of this book *only* we will say that for every 100 yards distance the object's *apparent* size to the naked eye is 10 percent smaller. There are a number of factors which influence the apparent size of an animal or object other than distance, such as shadows or whether there are other objects in between the object and the viewer. However, we are not

Photo taken by the author using a 50 mm lens, the standard lens used by most amateur photographers. The 50 mm lens closely reproduces what the unaided eye sees. Note the dark tree on the ridge in the background.

The tree in the above photograph was taken with a 35 mm camera mounted on a Bausch & Lomb four-inch reflecting telescope. This is the effect of power: the object is brought closer to the viewer.

tackling the subject of range estimating here. For our object with its decreasing size we can organize a table of distance/size (apparent) that would look like this:

100 yards	43.2 inches or 3.6 feet
200 yards	38.4 inches or 3.2 feet
300 yards	33.6 inches or 2.8 feet
400 yards	28.8 inches or 2.4 feet
500 yards	24 inches or 2 feet

Obviously the further an object is from the eye the smaller it appears to the eye. Optical instruments then, be it binoculars, scopes, or telescopes, are intended to reduce the distance between the eye and the object through magnification. To understand the results of power or magnification let's take our table and, using magnification in powers of 1.5×, 2×, and 4×, see the results.

Distance	Apparent Height (eye)	1.5× in./ft.	2× in./ft.	4× in./ft.
100 yds	43.2/3.6	64.8/5.4	86.4/7.2	172.8/14.4
200 yds	38.4/3.2	57.6/4.8	76.8/6.4	153.6/12.8
300 yds	33.6/2.8	50.4/4.2	67.2/5.6	134.4/11.2
400 yds	28.8/2.4	43.2/3.6	57.6/4.8	115.2/9.6
500 yds	24/2	36/3	48/4	96/8

At first glance this table appears to be wrong. How can a deer be the size of a horse? The truth is the object is only appearing larger through magnification, and depending on the object's distance, and the field of view of the instrument, the object will be either all in view or only in part. What has happened is that through magnification the apparent distance between the the object and the eye has been reduced. Nothing is actually larger nor is it any closer. Because everything viewed at

Fig. 6.1 .When you look at an object through a telescope with any magnification the object is enlarged (in this case four times) because the distance from the eye to the object is reduced.

once is enlarged the same amount the image seen will appear normal within the magnification (fig. 6.1) However, too much power can reduce the effectiveness of an optical instrument for a variety of reasons.

There are several factors which come into force when the power of an optical instrument is increased. Each one of these factors will result in a *reduction* in the instrument's effectiveness.

1. field of view is decreased
2. eye relief is decreased
3. exit pupil brightness is decreased
4. tremor is increased
5. resolution is reduced

In order to understand why increased power is often more

of a handicap than a help we must review each of these five factors.

Effects of Increased Magnification

When the power of an instrument is increased its true field of view will be reduced proportionally. The loss of this field of view will take place as power is increased regardless of the size of the objective lens. Consider a variable power scope of 3×–9× with a 40 mm objective lens. At 100 yards the field of view is 36 feet when the instrument is set on 3× (power), however when the power is cranked up to 9× the field of view is only 12 feet at 100 yards.

The reason for this is that magnification is the result of the focal length of the objective lens divided by the focal length of the eyepiece. Another way to describe it is that the power of the magnification is the ratio of the objective lens aperture to the aperture of the exit pupil. Since we know that the exit pupil size is the diameter of the objective lens divided by the power of the instrument we can actually see how the increase in power affects the exit pupil. To chart this relationship we'll use the three popular powers— 7×, 8×, and 10×—and couple them to a 50 mm objective lens.

50 mm Objective Lens			
Power	7×	8×	10×
Exit Pupil	7.14 mm	6.25 mm	5 mm

Consider what happens when you have an instrument with a 35 mm objective lens and you use the same powers of magnification.

35 mm Objective Lens			
Power	7×	8×	10×
Exit Pupil	5 mm	4.38 mm	3.5 mm

Obviously the size of the exit pupil can drop dramatically, affecting the image you see in the instrument.

Another effect of increasing magnification is the loss of field of view. There are three rules which affect field of view and these are:

1. the effective size of the eyepiece or ocular lens; this is not the diameter of the lens but the working aperture of the lens;
2. the *eye relief* of the instrument (eye relief is the distance from the eyepiece lens of the instrument to the user's eye; see fig. 6.2); and
3. the magnification of the instrument.

Notice that the size of the instrument does not have any effect on the field of view while magnification does. The way these three rules affect field of view begins with the first two rules working together. The eyepiece aperture and the eye relief or distance from the eye establish an *angle of view*, which is measured from the center of the pupil to the edge of the aperture in the ocular lens. This angle of view is then divided by the actual magnification to determine the field of view for the instrument. To increase the field of view of the instrument either the eye relief must be shortened or the size of the eyepiece must be increased. This problem is one that plagues the manufacturers of rifle scopes because the eye relief on a rifle scope must be great

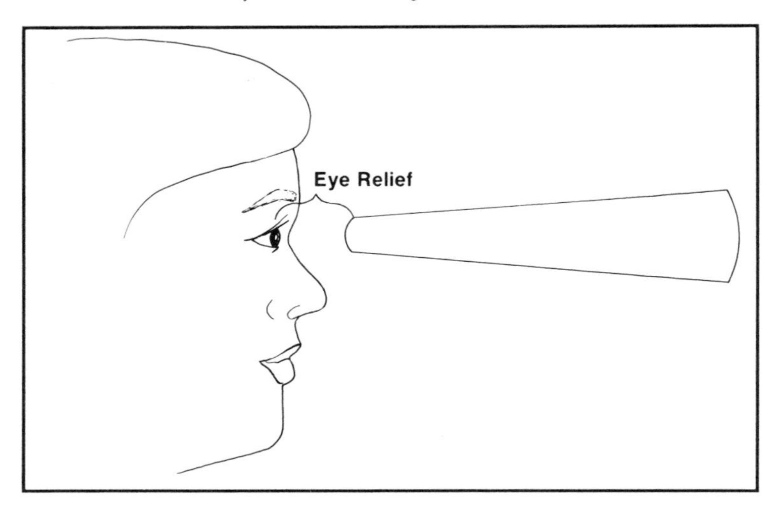

Fig. 6.2. Eye relief is the distance from the eye to the eyepiece lens in an optical instrument and is affected by the magnification of the instrument.

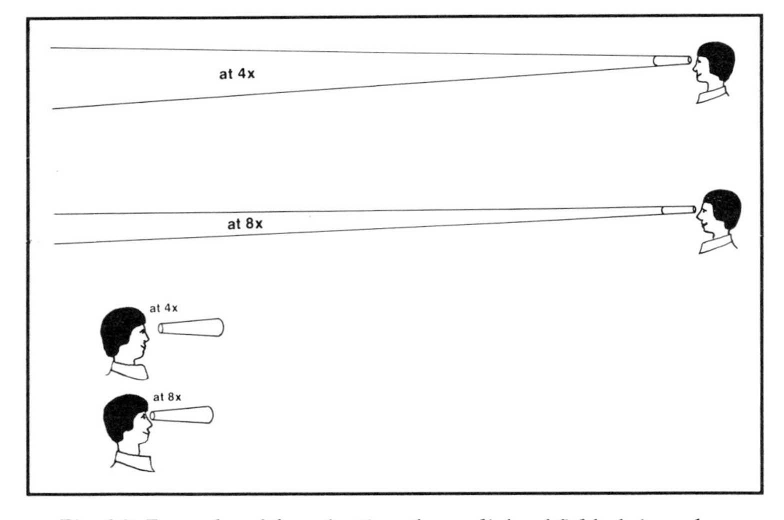

Fig. 6.3. Examples of the reduction of eye relief and field of view when the magnification of an instrument is increased

enough to allow safe handling of the rifle. If the eye relief is too short the scope can hit the user during recoil. If the eye relief is too great then field of view is decreased.

Magnification in this area becomes important because the angle of view is divided by the magnification. The higher the magnification number the lower the field of view and the less area you can see through the scope (fig. 6.3).

We have already covered exit pupil (EP) in chapter five, and we know that EP is arrived at by dividing the diameter of the objective lens by the power. If we begin increasing the magnification of an instrument without the corresponding increase in the size of the objective lens (which rapidly becomes a very expensive and impractical prospect), then the EP becomes so small that in some instruments, such as binoculars, the actual effectiveness of the instrument is lost. Some firms are responding to perceived demand in the marketplace for binoculars of 15× and are offering them in instruments with objective lens sizes of 50 mm or 35 mm. While the magnification might seem great at the time, consider the exit pupil size at 3.33 mm in the 50 mm instrument and 2.33 in the 35 mm instrument. Although the image may appear acceptable in full daylight, as the light begins to wane in the afternoon or, if you are trying to look into the shadows, the lack of a reasonable exit pupil and corresponding drop in relative brightness and DVI will result in your being unable to use the instrument for more than a few hours of the day.

Another problem of magnification is tremor. The higher the magnification of any optical instrument the more difficult it is to hold the instrument steady. And the higher the magnification the greater the exaggeration of your own movements. In binoculars any magnification over 10× is useless without some type of support to hold the instrument steady.

We must also understand why it is possible to buy a telescope, with its capacity for much greater magnification, only to discover that the instrument seems to be a failure or does not live up to our expectations. The reason is that there is a theoretical limit to the usefulness of magnification, which can be found using basic math; this limit is based on the fact that the average human eye can resolve detail down to about 60 seconds (or 1 minute) of angle. A second of angle is 1/60 of a minute of angle and a minute of angle is 1/60 of a degree of angle and there are 360 degrees in a circle. To give you some idea of the concept, a minute of angle is equal to one inch at 100 yards. In other words, we might say (although not accurately) that the average person can, with the unaided eye, differentiate between two objects separated by one inch at 100 yards.

Once the idea of a limit to what a person can see is understood, the concept of the theoretical limit of resolution of an optical instrument can be understood. The theoretical limit of an objective lens is the constant 114.3 divided by the diameter of the objective lens in millimeters. The result is the *seconds of angle* (SOA) of resolving power for that objective lens. Because the objective lens must be coupled to an eyepiece or ocular lens, which in fact enlarges (magnifies) the image formed by the objective lens, we can find the limit of the power of the eyepiece by dividing the resolving power of the human eye (60 SOA) by the seconds of angle of the objective lens. The result of this is the magnification limit of that objective lens.

This formula is applied to the use of refracting telescopes and explains, in part, why a high-powered telescope seems to "lose" its ability to provide a clear image when the highest power is used. Because of demand by the consumer most firms manufacture eyepieces for telescopes which actually exceed the resolving power of the objective lens. Consider the popular spotting scopes on the market with zoom lenses of 20× to 45× that are on 60 mm scopes. Since a 60 mm objective lens size is approaching the limit of what can be economically manufactured for the mass market (see chapter seven), it is consumer demand that has forced companies to manufacture these eyepieces, yet they actually exceed the capabilities of the objective lens. By using the formula just covered we can find the theoretical limit of resolution of the scope.

objective lens diameter = 60 mm

$$\frac{114.3}{60mm} = 1.905 \text{ SOA}$$

1.905 = objective lens resolving power

$$\frac{60 \text{ SOA}}{1.905 \text{ SOA}} = 31.5 \text{ maximum magnification}$$

With a maximum magnification of 31.5 the loss of detail coupled with the other problems of excess magnification covered earlier become apparent. As the power is increased these problems increase proportionally. Another point to consider is that any irregularities in the air itself will also be magnified a corresponding amount. Heat waves shimmering off the surface of the ground, dust particles, moisture, etc. will also be magnified as they appear in the field of view resulting in even more loss of detail.

The theoretical framework of optics consists of many for-

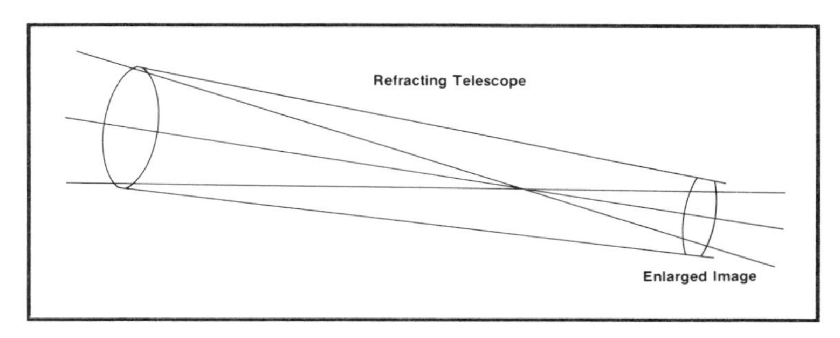

Fig. 6.4. An example of a simple refracting telescope.

mulas which are used to arrive at optimum lens sizes and shapes for various instruments. We cannot concern ourselves with all of these formulas and theories in a book on the practical application of optics to general observation and uses. However, if the principles I've covered are understood, anyone who is serious about using optical equipment in the course of their work or hobby—even though not trained in optical theory or a related field—can make an intelligent decision on what equipment they need.

In the following chapters we'll cover the types of optical equipment commonly used and discuss how they work and what their applications are. Even a limited understanding of the physics behind the equipment will reduce the mystery of optics and help you to make an intelligent investment in the equipment you need.

7
Telescopes and Spotting Scopes

North of our home is a bald ridge that gives a commanding view of the surrounding river valley. Sometimes, on lazy afternoons or in the evening, I set up a telescope and look at the ridge. On occasion I will see a hawk sitting in one of the few trees and sometimes a deer will amble along the ridge. For that reason there is always a telescope handy. Telescopes, binoculars, microscopes, cameras, and the outdoors have always been a part of my world. While I was growing up my parents encouraged my interest in "seeing and watching" whenever their limited budget would allow.

Over the years I have been grateful for that encouragement. At various times I have watched bald eagles feeding their young, pronghorn does teaching their spring fawns how to recognize danger, and from my own office I have watched a variety of birds in our cherry and apple trees. Optical equipment is a vital part of my outdoor experience.

Spotting scopes, or more accurately, sporting telescopes, are popular among today's outdoor enthusiasts. I've used my Bushnell Spacemaster spotting scope to scan distant peaks for deer and elk, to look at the rings of Saturn, and to watch pronghorns play. If there is any tool in my outdoor equipment inventory that I consider more important than my optical equipment it would have to be my knife. A top quality spotting scope can open a world of understanding by gaining you a glimpse into the lives of so many wildlife species, and it can provide hours of free entertainment if you have any interest in the stars.

Not every telescope qualifies as a spotting scope yet every spotting scope is, in fact, a telescope. There are many different types of telescopes; some of them are quite specialized in their uses while others, because of their construction and design, can be used for both recreation and serious observing uses. To pick out the type of telescope you need you should be aware of the different types available.

Refracting and Reflecting Telescopes

Galileo applied the principles of refraction of light through glass to build his telescopes. The idea of refraction was not new, but the application was. What refraction refers to is the fact that light waves, on entering any medium from "free space," are slowed down. There is a very specific amount of slowing down and in that process the waves of light appear to be bent. The classic example of this is, of course, the pencil in the glass of water (see fig. 3.4). The pencil appears to be bent because the light waves are bent by the water. This bending of light is the refraction and a *refractive index* tells exactly what effect various substances will have on light waves. By knowing the precise angle at which light will be striking the object, and knowing the substance's refractive index, the angle at which the light wave will leave the substance can be calculated. There is a simple law which covers this: Snell's Law states that the refractive index is the ratio of the sine of the angle of incidence (entry) to the sine of the angle of refraction (exit). *Refracting telescopes,* then, rely on this law to produce their images (fig. 6.4). Light waves are

Two examples of refracting telescopes which can be used for either terrestrial or astronomical viewing. Scope on the left is the author's Simmons 60 mm × 400 mm telescope; on the right is a Tasco 50 mm × 600 mm telescope. Both instruments are equipped with finder scopes and star prisms for use in astronomical viewing.

gathered by the objective lens of the instrument and are bent or focused on a specific point to produce an image that is once again picked up by a lens (the eyepiece) and focused. This second image is an enlarged image that is picked up by the eye (which also uses refraction). In a simple telescope without an erecting or neutral lens the image will be seen upside down by the observer.

Refracting telescopes are ideal for most recreational viewing (i.e. observation of wildlife and other nonastronomical purposes) and except for instruments made for scientific research few refracting telescopes are in use by researchers. A key reason is that it is impossible to produce an image in a refracting telescope that is absolutely colorless. Also, the requirements for

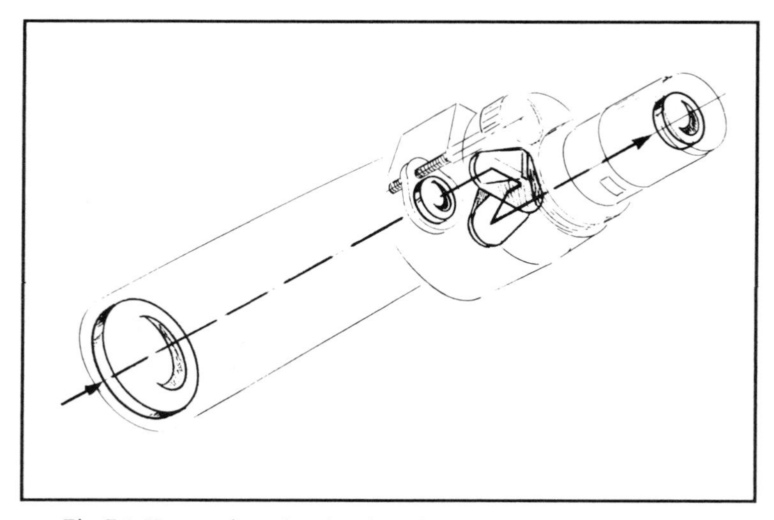

Fig. 7.1. How a prismatic refracting telescope works. Prism shortens the length of the instrument. (Illustration provided by Bushnell)

the glass make building telescopes with very large objective lenses impractical. The world's largest refracting telescope is the 40-inch refracting scope at the Yerkes Observatory. But for the average person who is interested in a telescope that is rugged enough to handle the bangs and bumps of being hauled around outdoors or set up for the study of wildlife under every possible weather condition, the refracting telescope is the obvious choice and will produce the best possible results in a reasonable cost range.

There is a reason for this preference and it primarily has to do with durability. In modern refracting telescopes, of the type we are concerned with, and within the tolerances of resolution of detail, refracting telescopes within the range of 40 mm to 75 mm are both economical and durable. The objective lenses and other elements within the scope can be mounted securely in the instrument so that they can stand up to a reasonable amount of abuse and still deliver an acceptable image within the tolerances of resolution.

A refracting telescope uses a large objective lens to collect the light and an eyepiece lens to perform the actual magnification. The most common type of refracting telescope is the amateur observing scope found in most department stores. These telescopes are impressive with their long tubes and array of interchangeable eyepiece lenses. Most refracting telescopes have a long focal length for the objective lens and in order to make use of this focal length a long tube or body is required. The reason for this is that when light is collected by the objective lens and is focused down the tube the refraction is such that it often requires a distance of several hundred millimeters to reach its focal point. Slightly beyond this focal point the eyepiece lens, which is much smaller than the objective lens, picks up the image and again focuses the light rays and enlarges the image. For my own casual star observing I frequently use a Simmons refracting telescope, and when friends come over for an evening of star watching I set up my Simmons as well as my Bausch & Lomb reflecting telescope.

If you are considering a telescope for casual observing of either the stars and moon or terrestrial objects around the home a telescope such as this is fine, providing you are willing to invest in a good one. Unfortunately this is also the easiest type of scope to build cheaply. A refracting telescope with a 40 mm objective lens and 2000 mm focal length sounds good and looks good until you realize that the long focal length alone will require a very stable mounting system that will insure the entire length of the tube is held rigid. Further, if you remember the resolution limits from the previous chapter, it is not going to give you any additional resolving power. If we apply the formula from the previous chapter we discover that the resolving power of the 40 mm objective lens gives us a final magnification of 20.98 or 21×. Any magnification beyond that is empty and serves no useful purpose.

I use the 40 mm objective lens size in this example because this is one of the most common. In looking over one large chain store's advertisements I found that they offered a complete telescope for observing Halley's comet during its pass that had a 1.5-inch objective lens and a power of 40×. If we convert the 1.5-inch lens to millimeters we have 38.1 mm and if we apply the formula for finding resolution we discover that the magnification limit of this lens is 20×, *providing the optical glass is any good.* In short, the telescope would make a handy toy but for anything you wanted to see beyond 20× it would be virtually impossible to get a clear view. There is also the consideration that this magnification would severely limit the field of view as well.

My purpose here is to persuade you not to buy a telescope based on some discount store's advertising hype about an instrument's power. There is an additional way to design refracting telescopes which enables manufacturers to build larger objective lenses (which allows for greater magnification) and, by the use of prisms, shorten the overall length of the instrument without a corresponding loss of focal length. This type of refracting telescope is the *prismatic refracting scope* (fig. 7.2). These telescopes have several important advantages over the straight refracting scope. First they are more compact, being as much as two-thirds shorter, and they are more solidly built. By using a prism to reflect the light rays in the instrument several times, the focal length is not altered and the larger lens of 60 mm or more is possible in a compact instrument. Many of these scopes offer magnifications above the resolving power of the objective lens, however. Again, this is in answer to the demands of the consumer; such scopes are usually limited in use to the viewing of set objects, such as distant mountains. The image seen is not crisp and it is often only marginally acceptable.

For most outdoor uses the prismatic refracting telescope is superior to any other and it is useful for amateur star gazing as well. It is also obvious that the amount of light collected and the image seen is somewhat limited. The size of such scopes is also limited because of cost. Going beyond 60 mm the cost of the objective lens becomes prohibitive for the average person's pocketbook. There is, however, an alternative and that is the *reflecting telescope.*

Reflecting Telescopes

Reflecting telescopes use a mirror to collect light and focus this light where it can be picked up and magnified by the eyepiece lens. The advantage of this type of scope over the refracting scope is that a much larger aperture can be used. Rather than the objective being limited to 60 mm for a reasonably priced commercial telescope, reflecting telescopes can be within the average person's budget while equipped with six, ten and even fourteen-inch mirrors. For scientific viewing, reflecting telescopes can be built with mirrors hundreds of inches in diameter. For our purposes we'll concern ourselves with more practical dimensions such as four inches and up.

Reflecting telescopes come in a variety of designs such as the Newtonian, the Cassegrain, and the Gregorian. All of these designs rely on the principle of reflection to focus the light rays where they can be picked up by the eyepiece. In the Newtonian design the primary image formed by the mirror is diverted by a small, flat prism mirror out the side of the instrument and into the magnifying lens (fig. 7.3). In the Cassegrain design a convex secondary mirror about a fourth the diameter of the primary mirror reflects light back through a hole in the center of the primary mirror which multiplies the focal length of the instrument giving the effect of a very long focal length in a very

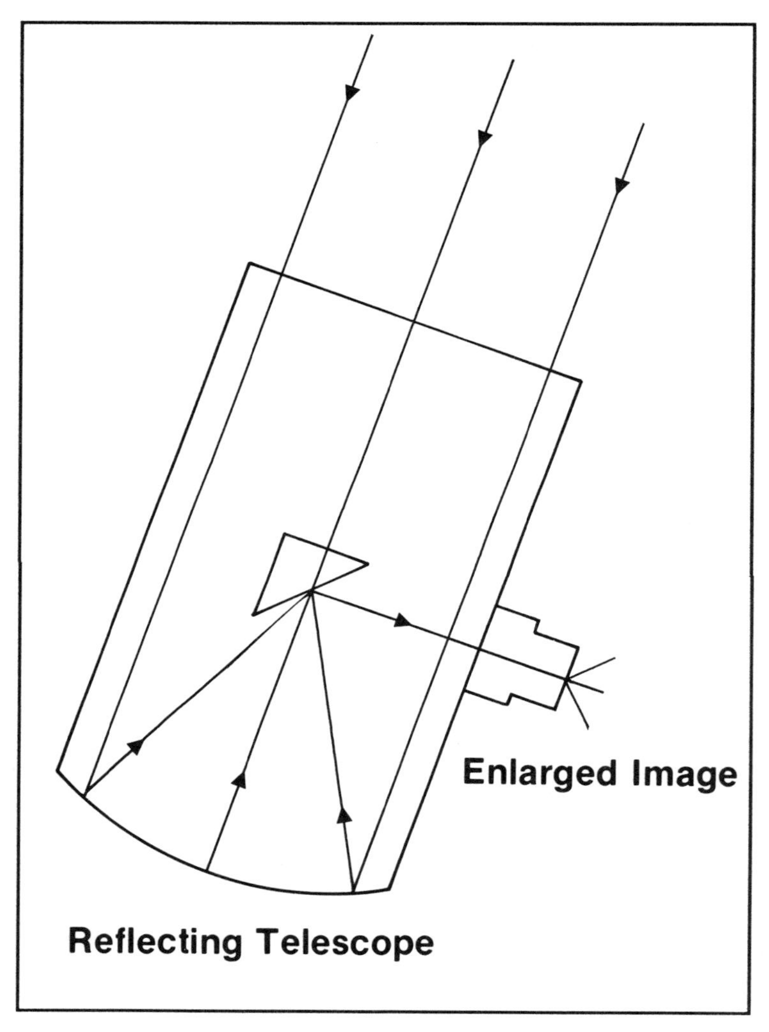

Fig. 7.2. An example of a Newtonion reflecting telescope using a prism mirror to reflect light collected by the mirror into the eyepiece of the instrument.

The author's Bausch & Lomb Criterian 4000 four-inch reflecting telescope. Although this instrument can be used as a terrestrial spotting scope its primary use is as an astronomical viewing instrument.

compact instrument. The result is that more magnification can be employed in the instrument.

Reflecting telescopes have an advantage over refracting scopes in that more magnification and clearer images can be obtained at a fraction of the cost of a refracting instrument. Their single most depressing drawback is that they are fragile. The precise alignment of the mirrors to the eyepiece requires constant care. However, reflecting telescopes are so much more compact and lighter weight that they are superior to the lower-power refracting telescopes in this respect alone. As for image

The author's Bausch & Lomb four-inch reflecting telescope showing the front of the instrument where light is collected and passed down the tube to the mirror in the base, where it is then reflected back toward the front before being reflected again and focused for the eyepiece to enlarge.

detail, a good and well-cared-for four-inch reflecting telescopes will deliver several times the clarity of image at higher magnification than a commercial refracting scope of less magnification.

Catadioptric Scopes

Within the reflecting telescope design is the catadioptric design of telescope (fig. 7.4). This more recent design is a type of spotting scope incorporating both the reflection and refraction principles in one scope. The result is a very compact instrument that has the durability of a refracting scope and which, by using

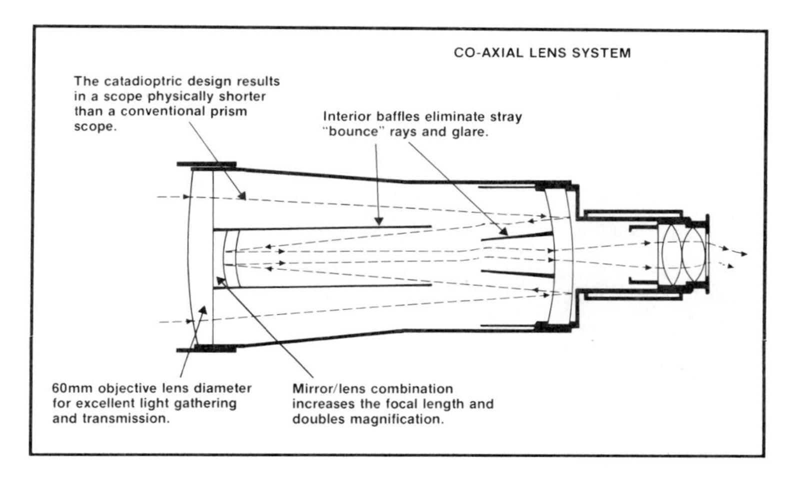

Fig. 7.3. How a catadioptric telescope works. This illustration is taken from the design used by Redfield® Optics of Denver, Colorado.

reflection principles, can make use of a very long focal length. These scopes can employ a 60 mm objective lens for greater light-gathering capability while remaining very compact. As part of a field test one summer I used a Redfield catadioptic scope in the mountains while researching magazine articles. The ability of the scope to handle the rough and tumble abuse proved the principle of combining both types into a single scope for a lightweight and very durable scope for outdoor use. Another advantage of this type of scope is that the greater light-gathering capability of the larger objective lens allows for more magnification in the eyepiece.

Which Scope to Use and Buy

The decision to buy a particular type of telescope should be dictated by the intended use of the scope and not by its design.

Bausch & Lomb 15-60× Discoverer Zoom Telescope, Model 78-1600. (Photo by Bushnell)

If your interest leans more toward packing up your scope and taking to the hills to do your observing of wildlife, with only casual interest in celestial or backyard observing, then a prismatic telescope with a 60 mm objective lens is the logical choice. A variable magnification up to 30× will enable you to easily make use of the scope under a variety of conditions. The same type of scope can be used in the field for spotting game and is rugged enough to stand up to being hauled around in a Jeep or by horseback if it is carried in a reasonably good carrying case.

In recent years several firms have taken to adding hard rubber armoring around spotting scopes to make them even more rugged. Most armored scopes drop down to an objective lens size of 40 mm to reduce the overall weight of the instrument. These armored scopes are excellent pack scopes and can stand up to an even greater amount of abuse in the field. Their range of use is more limited and while they will still enhance

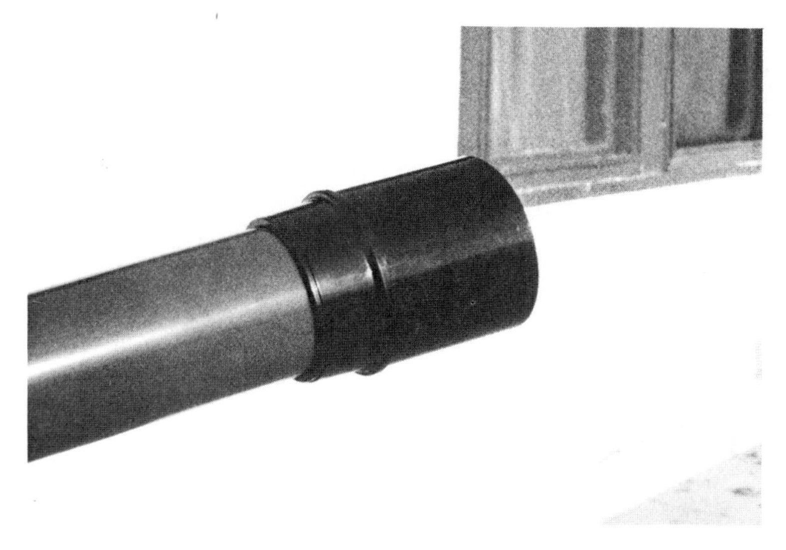

Example of a dew cap used to keep moisture off a telescope's objective lens during night viewing.

your viewing activities the amount of magnification is reduced to 20×.

Another important consideration when choosing between the 40 mm and 60 mm is time of day. The light-gathering capability of the 60 mm lens makes it more convenient for use in late hours. The same can also be said for wildlife observing. Since most wildlife species are active in the early morning or early evening hours the additional 20 mm of lens size does increase the light-gathering capability. If your primary intent, however, is to use it in the field while hiking or climbing, the extra protection afforded by the rubber armoring can be an insurance policy that you won't want to give up. If this is your primary use, the solution is to reduce the magnification so that your image is brighter. It is actually easier to observe an animal's

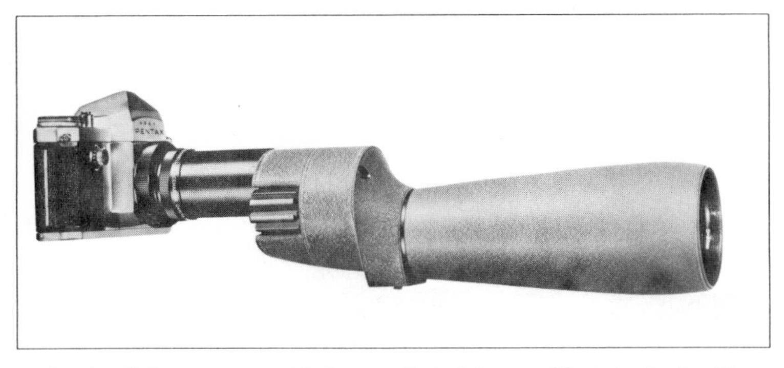

Bushnell Spacemaster with Camera Body Adapter. (Photo by Bushnell)

activities by using a lower power on the scope that produces a brighter and clearer image than by using a higher power that produces a dimmed image.

Stationary observing from a blind allows you to move up the scale of telescope types to the more powerful and effective reflecting telescopes. If you are observing wildlife from a blind, or any other situation where you are not going to be hauling it around but, in fact, will be setting up the scope where it is protected from the weather, and you need both magnification and resolving power, the obvious choice is a reflecting telescope. Depending on the diameter of the mirror, you can enjoy magnifications of 60×, 100×, and higher. A four-inch reflecting telescope equipped with a camera body can give you the effect of a 4000 mm focal length lens! Also, because the mirror is gathering more than six times the amount of light as a 60 mm lens you can use a reflecting telescope in low-light situations where a standard refracting or prismatic refracting telescope would be nearly useless.

When you are looking at a scope check to be sure it can be mounted on a standard photographic tripod and that when it is mounted the scope is not subject to shudder. I've used several

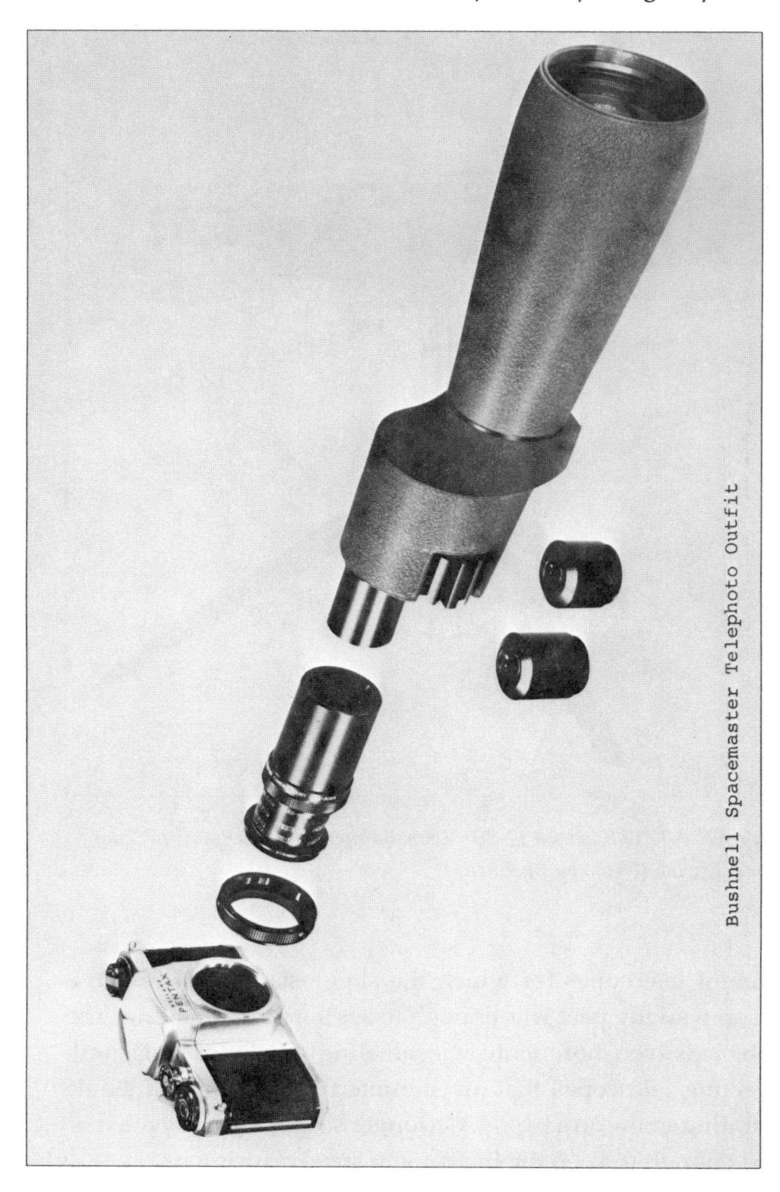

Bushnell Spacemaster Telephoto Outfit. (Photo by Bushnell)

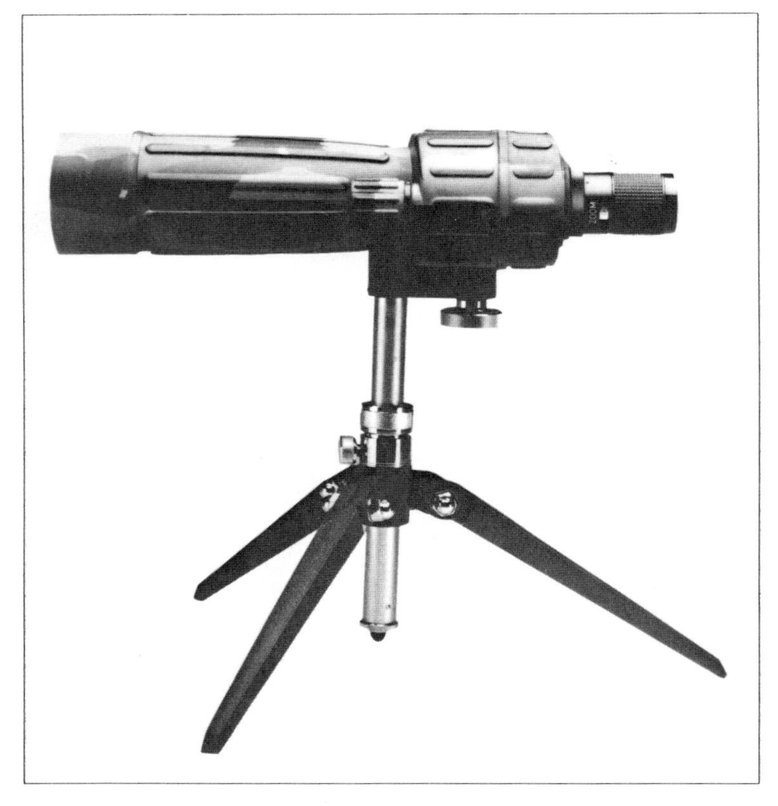

Model 78-1717 Bushnell 12-36× Zoom Banner® Trophy Spotting Scope and Tripod. (Photo by Bushnell)

different telescopes for which the slightest tremor of even a person walking past was enough to set the scope shaking. The problem is even more acute when dealing with long focal length refracting telescopes that are mounted by means of a single mounting screw on a tripod. On longer scopes the body must be rigid enough to keep the instrument steady during use.

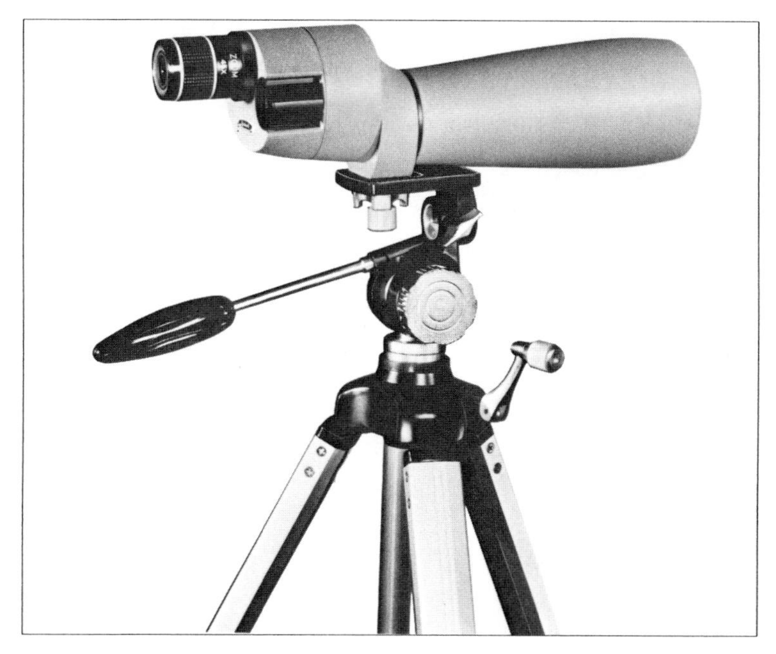

Bushnell Zoom Spacemaster, Model 78-1809, on All Purpose Tripod, Model 78-3101. (Photo by Bushnell)

The following recommendation table gives you some idea of what to look for when you want to buy a telescope.

Purpose	Type and Objective Lens Size
Recreational wildlife viewing	40 mm with 20× magnification, refracting, or catadioptric
Field study of wildlife	60 mm with 30× magnification, refracting or catadioptric
Stationary viewing	4-6 inch reflecting

When you go into a store to start looking at telescopes keep these general recommendations in mind and be sure you test every scope you look at. If it is at all possible take the scope outside and use it under normal sunlight conditions and look for the problem areas discussed in previous chapters. You should know by now what to expect out of your telescope's optics. If it doesn't perform as it should for the price, or is an obvious hype job, don't waste your money.

8
Binoculars

In the predawn light we were gulping down the last bits of a breakfast consisting of sweet rolls and hot coffee between attempts to collect our gear for a day of stomping around in southern Colorado's Sangre de Cristo mountains, when one of our clients picked up my Bausch & Lomb Discoverer binoculars and turned them over.

"You carry these?" he asked.

"Sure do and wouldn't trade them for anything," I answered then returned to my problem of jamming lunch into a backpack.

"What power are they?"

"Those are seven by twenty-fours," I explained.

"Not powerful enough for me," he quipped then lifted his own binoculars, an oversized pair of 7×50's without benefit of any armoring. They were typical department store binoculars. I started to explain to him that in the first place my compacts were just as "powerful" as the monsters he was holding and had superior lenses, coatings, and a host of other refinements that made them superior, then thought better of it. After all, he was a paying client and I didn't want to offend him since he had paid his tab in advance. I was also lucky enough not to have to accompany him that day so I wasn't going to have to listen to him complain later in the day about his binocular's weight, or if it started raining or snowing, about them getting all fogged up—both of which occurred before afternoon so Tim had to listen to his complaints.

Many myths about optical instruments used in the outdoors have grown to be accepted as fact though they are nothing more than a misunderstanding of what is really happening. By now you should understand optical equipment well enough to appreciate the fact that both my compacts and his "hernia glasses" had the same power. Also, it should be evident that a good portion of an instrument's capability is directly related to the quality of the engineering behind it. As long as the manufacturer has not attempted to "enhance" the objective lens with too much power, the resolving power of the instrument is such that you will still receive a clear image. If the optical glass elements are welded together properly, and the coatings on the glass are all within closely controlled tolerances, the overall effect is a quality instrument. My point is that size is not a determining factor in any binocular's power or effectiveness but size is too often an excuse for building a marginal instrument at a lower dollar figure where the lack of quality is hidden under the size of the objective lens. In the case of quality instruments size is a controlled engineering function intended to deliver the finest possible performance under extended and marginal conditions; in other words, the manufacturer is not trying to hide faults but is using the size to bring out more of the positive effects of a larger instrument.

A 10×50 Swarovski Habicht compared to a 10×50 discount store instrument is about the same as comparing a Jaguar to a Ford. Both are automobiles. Both will get you from point A to point B. One will make the tight curves and passing power

This binocular is a perfect example of what an inexpensive instrument can do—fall apart! The left eyepiece broke off and both objective lens elements separated from the body of the instrument.

seem smooth and a joy; the other will simply accomplish the task required but "something" will be missing. The choice is part yours and part your pocketbook. Between the two extremes, however, there is a parade of quality automobiles that will move you up the scale of reliability, form, and function. The same holds for binoculars.

What Are Binoculars?

So far I have used telescopes of various types to explain the different faults to be found in optical instruments. My reasoning on this is that *binoculars are simply two telescopes on parallel lines.* There is, then, one telescope for each eye. Prisms are used to shorten the length of the instrument while maintaining the focal length required for the objective lens. The prism also serves to

Stereoscopic vision is produced by the distance between the two objective lenses. Although instruments with less distance than these 50 mm binoculars may be easier to handle the actual distance they can allow the user to see is not much further than the unaided eye.

reduce the number of elements needed by reversing the image and presenting it to the eye so it will be seen upright rather than upside down and backwards, as would normally occur in a simple refracting telescope.

An interesting feature of binoculars is their effect on *binocular vision* (also called stereoscopic vision). This is the ability of the human brain to fuse the two fields of vision seen by each eye into a single image. The human eye is not able to produce this vision if the angle at the eye from the object is less than 1/120 of a degree. The distance of separation of the eyes averages about 65 mm and allows us a maximum distance for stereoscopic vision of around 444 meters. Binoculars, because of their magnification, increase the stereoscopic vision distance dramatically. Seven power binoculars, with a distance between the objective lenses of 130 millimeters, will increase a person's stereoscopic vision range to more than 6,000 meters. This stereoscopic vision range will increase or decrease in proportion to the distance between the two objective lenses and their size. To be effective for long-distance viewing there must be a separation of the objective lenses.

If binoculars are nothing more than two small telescopes hooked together there must be a reason for the wide swings in prices, styles, and quality. There is, and it lies in the fact that binoculars are more adaptable to a wider variety of purposes than any other optical instrument bought by the public. The binoculars you might carry to a Friday night football game to unravel the mystery plays by your high school football coach are not the same instruments you would necessarily carry on a backpacking trip. The same is true, of course, for the binoculars carried into a theater: they would be useless at the football game! For the most part the common denominator of binoculars is that they are hinged telescopes with two or three focusing features. As such they also have special optical problems in addition to those covered in earlier chapters. We explore some of these special problems later in this chapter.

The Development of Binoculars

The first binocular instrument is believed to have been constructed in 1608 by Johann (Hans) Lippershey who, as I have already pointed out, is credited as one of the inventors of the early telescopes. Lippershey did not really do much with binoculars and the next reference we have is Capuchin Antonius Maria Schyrleaus de Reheita describing the construction of double terrestrial telescopes. Finally, Capuchin Cherubin d'Orleans followed with larger double telescopes which were equipped with adjustments for the interocular distance for individual users and eyepiece adjustments to allow the instruments to be focused on objects both near and far away.

Very little interest was taken in binocular instruments until about 1823 when the Viennese optician Johann Voigtlaender received a patent for a binocular instrument having two telescopes whose axes were parallel and in line with the center between the two. In 1825 a Frenchman, J. P. Lemiére, obtained a French patent for an instrument with a common central focusing adjustment and the ability to change the interocular distance by turning the two barrels around their mechanical axis. The last apparent improvement to take place for some time was made by another Frenchman, P. G. Bardout, who combined the other improvements with a central focusing arrangement for both eyes, which allowed him to obtain higher magnification.

The invention which really gave binoculars their start however, was by Ignazio Porro who in 1851 invented a prism combination which is named after him—the *Porro prism*. The Porro prism consists of two right-angle prisms placed so their hypotenuse faces are adjacent and their faces of total reflection are at right angles to each other. The effect is the complete reversal of the image so that the reversal caused by the objective and ocular lenses is reversed back and the image seen by the eye is upright.

Again, a Frenchman was first to make use of this prism in a binoculars: A. A. Boulanger introduced a binocular instrument that both made use of the Porro prism and was equipped with screws to change the interocular distance. Boulanger did not recognize the possibility of improving the distance between the objective lenses; this improvement was made by C. Nachet in 1875.

Finally, binoculars began to evolve into today's instrument through the efforts of Ernst Abbe, starting in 1893. Abbe improved binoculars first by mounting the prism so that the distance between the objective lenses was increased. Then, by separating the two hypotenuse faces from each other, he was able to shorten the overall length of the instrument. With Abbe's improvements binoculars became a compact instrument.

There have been a number of variations of prism designs over the last century, however most of these have been of one of two types. The second prism design of importance is that by Giovanni B. Amici. This design, commonly called the *roof prism,* is today one of the most demanding to design and through modern engineering is appearing more commonly in small compact binoculars. Where the Porro prism is often large, even in modern instruments, the roof prism designed by Amici allows the bundles of rays traveling through the entrance and exit faces of the prism to be reflected symmetrically from one roof to another, with the result that the image is completely reversed so that, again, the objective and ocular lens effect of reversing the image is cancelled. The accuracy required in building a roof prism is such that the right angles must be within two seconds of arc to avoid doubling the image. There have been a number of improvements in the modern roof prism until today it is an economical and compact system used in binoculars.

The two types of prisms are both commonly used today in binoculars and are found in instruments having the same power but different design and use. Because the roof prism functions within the single body or tube of each barrel of the instrument the distance between the two objective lenses will be the same or very close to the interocular distance of the eyes. So while a roof prism instrument may have a seven power magnification, because of the distance between the objective eyepieces the stereoscopic vision produced by this instrument will be less than that for an instrument of the same power using a Porro prism with wider separation of the objective lenses.

There is an important consideration here about the uses of the two types of instruments. If you are scanning wide areas of terrain, or even a single object at a great distance, the farther apart the objective lenses the easier it will be for you to determine distance between two objects in the field of view. For viewing at distances of 350 to 400 yards, either instrument will be adequate for your purposes. A serious wildlife observer who uses optics in the course of his work, or even a serious amateur really should be equipped with two instruments—one for long-range viewing (over 250 yards) and another for closer viewing. I have found myself switching from one instrument to another because of the difference in ranges. The point to remember is that these small binoculars are most effective at shorter distances because of the limits of the stereoscopic effect. When you need binoculars for long-distance viewing you should opt for a larger instrument. This is not a license to grab for more power but simply a recognition of the requirements of stereoscopic vision and of how viewing can be improved and eyestrain reduced.

Collimation

You might want to think of *collimation* as a fancy word for alignment. Keeping in mind that binoculars are two telescopes mounted on a hinge, you will understand the first problem with binoculars, and that is keeping three axes aligned. The mechanical axis is the axis which extends through the center of the hinge holding the two barrels together. Each barrel or telescope then has its own "optical axis." Both sides of a binocular must be parallel to each other at all times in order for it to produce the stereoscopic vision of the eyes (fig. 8.1). To do this the barrels must be mounted so they are also parallel to the mechanical axis or the hinge of the binoculars. When either of the barrels or sides of a binocular is no longer parallel to the other's axis the

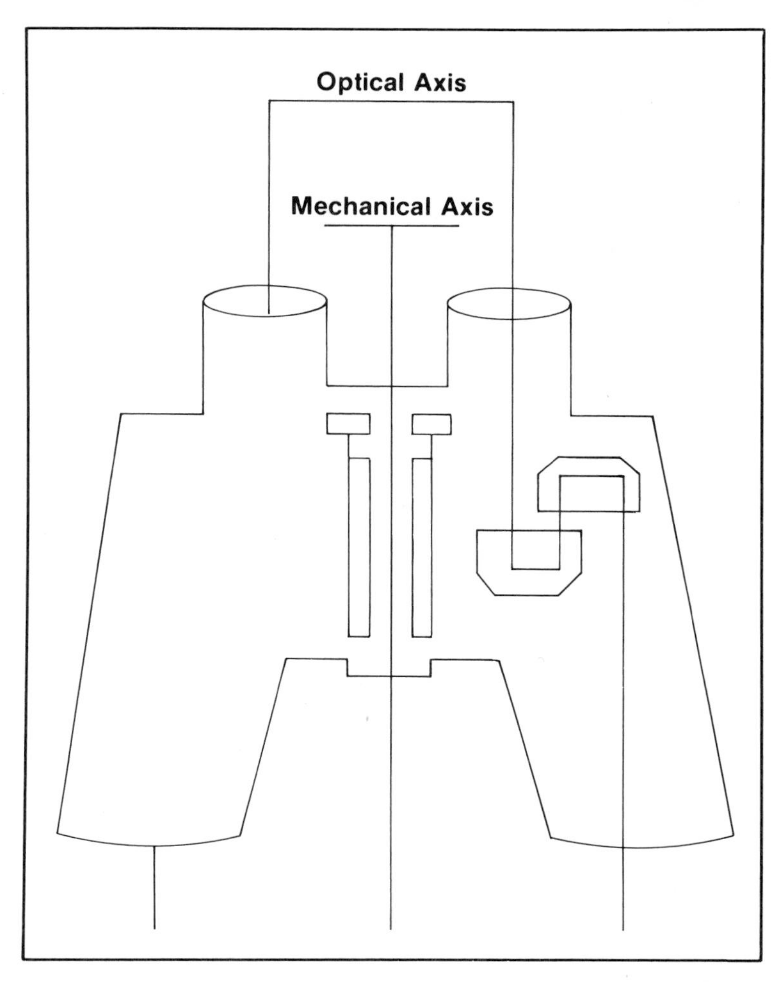

Fig. 8.1. The three axes of a binocular instrument. The optical axes must parallel the mechanical axis.

instrument is out of collimation; the image is no longer stereoscopic and will appear slightly blurred or doubled. This same effect can occur when a prism inside one of the barrels is

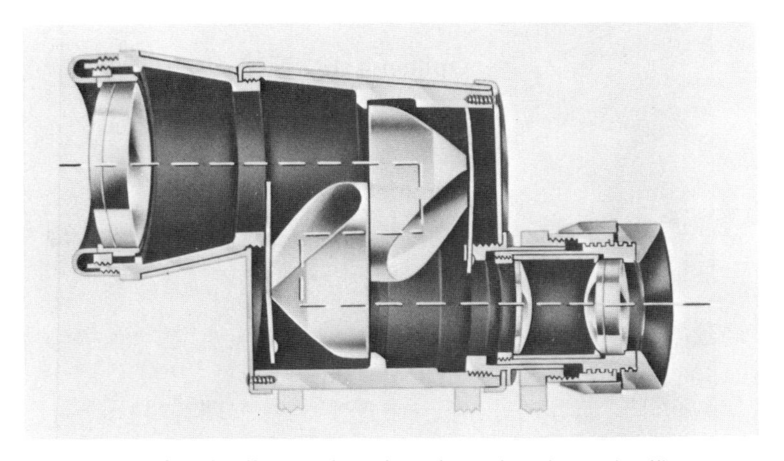

Cutaway of Bushnell Sportview Binocular. (Photo by Bushnell)

knocked out of alignment so it is no longer true to the optical axis of that side of the instrument.

Because of the variation in people (we just aren't all alike), all binoculars are equipped with a hinge that allows you to rotate the two halves of the instrument around the mechanical

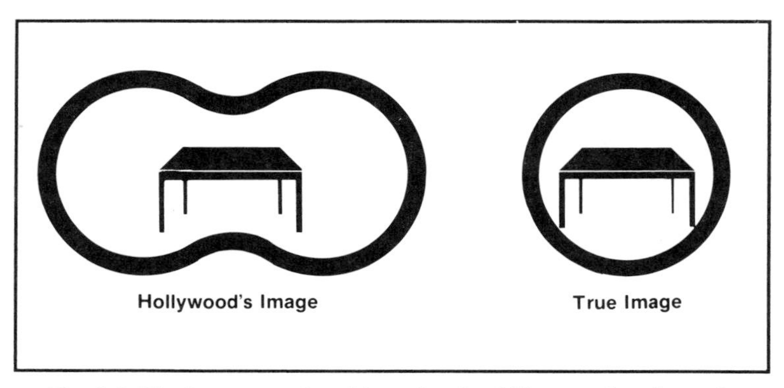

Fig. 8.2. The image seen in a binocular should be round and not the figure eight version used by Hollywood to give the viewer the feeling they are looking through binoculars.

On the left is the author's 7×35 Bushnell individual eyepiece focus binocular and on the left is a 7×35 Bushnell Instafocus® instrument.

axis until the two eyepieces or ocular lenses are properly adjusted for the distance between your eyes. This movement of the two halves is called the *interpupillary adjustment* (IPD) of the instrument.

Hollywood has created one of the myths that has resulted in too many people failing to make use of the hinge on binoculars to correct the IPD. The popular movie scene where someone is looking at a distant bad guy through binoculars shows a figure-eight outline of the image area (fig. 8.2). Remember that binoculars provide a stereoscopic image for the eyes. The eyes cannot get this stereoscopic image if the binoculars have not been adjusted until a single round image area is seen. Whenever you are using binoculars be sure you adjust the instrument to where it appears you are looking through only one eyepiece (although in fact you are looking through two).

For general recreation use the most popular design for the center hinge is joining the two halves on a shaft with the focusing wheel for the instrument at the rear of the shaft. The total amount of interpupillary adjustment available is somewhat limited because of the need to securely fasten the two barrels to the shaft so they are parallel. Behind the focusing wheel is another two-part hinge similar to those holding the main body of the instrument to the shaft and this is attached to the two eyepieces. As the focusing wheel is turned it moves the eyepiece in or out, by means of gears, as required to obtain focus. Both eyepieces must move together to obtain proper focus. If this hinge is bent or loose the result is unequal movement of the two eyepieces and the instrument will not provide a clear focus.

While not a great deal has changed in this construction over the years, improvements have been made and more binoculars are being manufactured with this shaft now a longer hinge that provides greater strength for the instrument. It is not uncommon to find an older binocular with a worn shaft or focus screw, or worse, with broken hinges for the eyepiece.

Exit Pupil and Prism

Sometimes it is difficult to convince a person that what they pay is what they get in optical equipment. Because optical equipment is always going to cost on the higher side of a dollar there is a natural tendency to reduce the investment, thinking there cannot be that much difference in equipment until you get into instruments costing more than $200. Nothing could be further from the truth. When someone asks me what they should buy for optical equipment my advice has been and remains get the best quality you can afford, foregoing the magic of power for clarity. To make up for inferior quality in optics some of the "bargain" brands opt for larger objective lenses (remember the amount of light entering the lens) so they can use

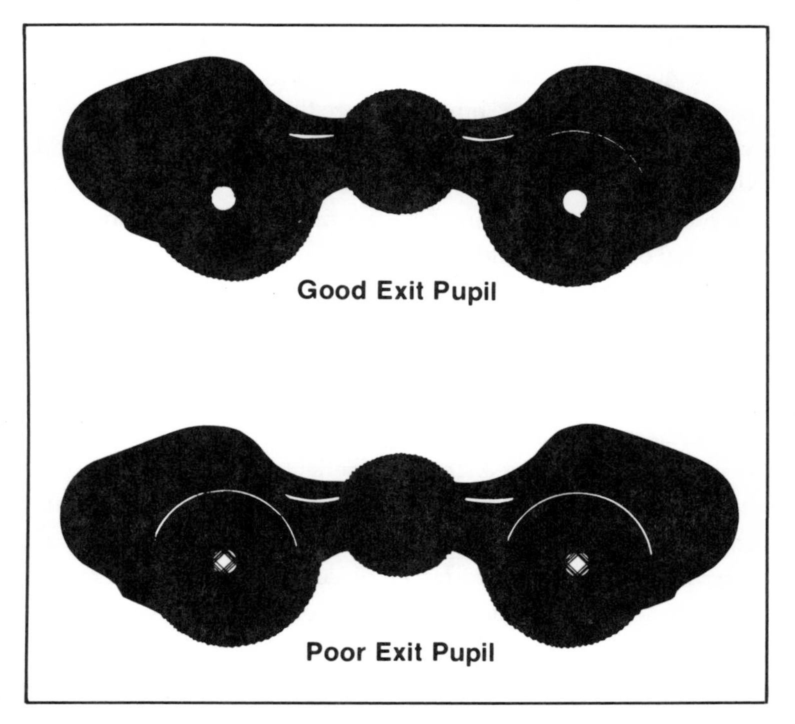

Fig. 8.3. By holding a binocular in front of your eyes so that you can see the circle of light produced in the eyepiece lens you can determine whether the instrument has a poor exit pupil or good exit pupil.

a poorer quality glass both in the lens system and the prism to produce what might appear to be a more economical yet quality product.

Other than watching for the faults I pointed out earlier in this book, there is another method you can use to check the quality of any binocular you are considering: look at the exit pupil in the eyepiece. If you don't remember what the exit pupil is, refer back to chapter five. The exit pupil on an optical instrument is the diameter of the light rays emanating from the eyepiece of the instrument. The size of the exit pupil is determined by dividing the objective lens size by the power of the instrument. Remember, binoculars are *two* telescopes mounted so that they are parallel to their own axes and the mechanical axis they are mounted on. And both sides are exactly equal so you only divide for the one (35÷7=5 in a 7×35 instrument). An exit pupil of 5 mm would then be seen in the eyepiece or ocular lens. You can see this exit pupil simply by holding a binocular about a foot in front of your eyes and moving it slowly around until you can see two circles (or circlelike) areas of light in the eyepiece.

A characteristic of binoculars is that the light must pass through the prism and each prism reflects the light two times. Since there are two prisms the light is reflected a total of four times. Again referring to earlier chapters you will recall that each time light passes through an object or is reflected there is a certain amount of scatter and absorption which is determined by the characteristics of that material. Because the light entering the prism is entering both straight (along the axis) and at angles, there is a tendency of that light which is not along the axis but on the edges of it and at an angle to the surface to be more easily scattered. By the time this effect is multiplied four times within a binocular (that is, four times in each barrel) when the light reaches the eyepiece it has a shadow effect. To compensate for this effect manufacturers must use a denser glass for the Porro prism so that it reflects the light rays in a more uniform pattern. The denser the glass the more uniform the exit pupil will appear. In "economy" binoculars this exit pupil will appear as anything from a bright diamond within a fuzzy diamond to a shaded circle of light. In better quality instruments the exit pupil will appear as a uniform circle of light with little or no shading (fig. 8.3).

This effect with the exit pupil is only found in binoculars

using a Porro prism and not in those using a roof prism design. You should also understand that the density of the glass in the prism has little to do with the overall quality of the instrument; a manufacturer that uses a Porro prism that produces this effect is certainly not going to invest any extra money in putting quality optical glass into a binocular that has a cheap prism. The effort would be lost by the light's scattering within the prism! In short, you get what you pay for.

Obviously very few of us can afford to invest $300 to $400 in the kind of quality instrument that produces exactly what we are looking for. The way around this is to recognize that you must invest your hard-earned money in an instrument that performs and is within your budget. There are good quality economy binoculars made by well-established companies. You *cannot* buy a 10×50 binocular for $75 that will be durable, free of most of the faults, and deliver a good even bundle of light rays to the eyepiece *and* be both water and fog resistant. You *can* buy a 7×35 instrument for around that dollar figure that is well made and performs. The secret is to compare by looking for the exit pupil and checking for the faults and other problems that the manufacturer needs to overcome. At that price they won't be eliminated but a competent manufacturer will reduce them to an acceptable range.

Image Tilt in Porro Prism Instruments

On occasion you will encounter a binocular which seems to have something wrong with it but you cannot quite figure out what. *Image tilt* is frequently found in instruments using a Porro prism and is a result of the two Porro prisms in either (or both) barrels being mounted so that the hypotenuse faces are not in alignment. This problem is somewhat common in older binoculars that have suffered through years of abuse.

To check for image tilt simply study a vertical line through

Bushnell 8×21 Sportview Folding Roof Prism Binocular, Model 13-8210. (Photo by Bushnell)

the instrument and see if it is appearing tilted in the image. If so the prisms are out of alignment.

Roof Prism Faults

The small, compact, and fairly durable roof prism binoculars are gaining popularity because they can be carried in your pocket and whipped out when needed. Remember that the roof prism requires a precise right angle that must be within two

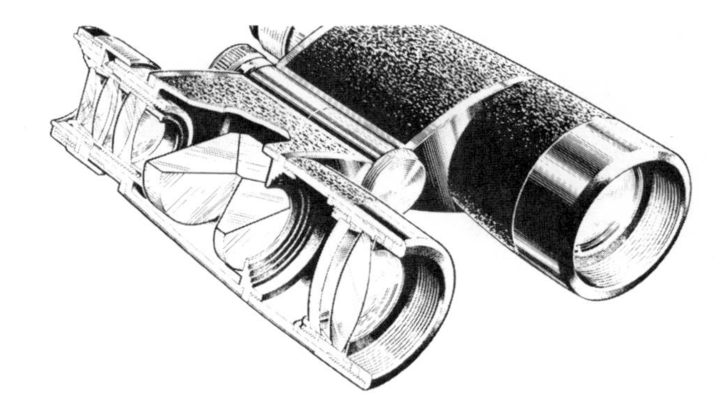

Cutaway of Bushnell Roof Prism Binocular. (Photo by Bushnell)

seconds of arc to be effective. The most common fault with these binoculars is that the prisms are not properly aligned and there is a slight doubling of the image. The way to check for this is to look at a single vertical line, such as a door, and watch for this doubling effect.

Another effect is a softening of the image in the eyepiece caused by imperfections within the prism. This occurs when the prism has small imperfections within it which cause a slight scattering of light rays along the axis of the light rays. In either case simple careful comparison shopping can eliminate the poor quality instruments.

Focusing

One of the major advances in binoculars design was the development of a method whereby both eyepieces could be focused at the same time by turning a single screw or knob. There are two areas of focus on binoculars. The first is the *main focus*

Bushnell's Instafocus® wedge replaces the focus screw and allows a closer reaction time to the eye's focus when using the instrument.

(usually in the center) which moves the two eyepieces back and forth in relation to the objective lenses. There are two focus possibilities, plus and minus. As you move an optical instrument closer to an object you must use plus focusing; the eyepiece must be moved outward or away from the objective lens. As you move further from the object you must use minus focusing or move the eyepiece closer to the objective lens.

I should note here that there are a variety of focusing devices on the market today. Some use a thumb screw, some a wheel, others have a wedgelike device replacing the wheel. The point to remember here is that when you focus binoculars you have a tendency to "over-focus." In other words, while looking at an object and attempting to get it in focus you go past the point where it is in focus, then attempt to return to that point. What is happening is that the eye and the fingers are not

The author's individual eyepiece focus binoculars shows the focus reference points used by the author.

working together. Of the various devices the least desirable is a thin thumb wheel set between the two barrels. The most comfortable focusing devices are the larger thumb wheels that allow fingers from both hands to rest comfortably on the focusing ring and work together, or the Insta-Focus® levers used on Bushnell instruments. The focusing action, regardless of the type, should be large and comfortable so that both hands can be working to focus the instrument. When you are checking out binoculars check the focusing mechanism. If you find yourself running back and forth trying to get in focus you are causing strain on your eyes and the instrument is poorly constructed.

The second area of focus is the *fine focus*, which should only need to be set once; it is usually only on one eyepiece. This allows you to focus that particular instrument to your eyes. There should be a 0 setting with gradations of + and - to each side

Bushnell's Zoom/Instafocus binoculars are popular for recreation viewing.

that are either numbered or color-coded so that you can remember where you set it in case you loan out your binoculars (not a good idea). As soon as you get your binoculars you should read the manufacturer's instructions and adjust them using the fine focus.

No Focus Binoculars

Next to my desk is a Bushnell individual eyepiece focus wide-angle binocular which I have used for several years and dragged through snow, rain, mud, and a host of other conditions. These binoculars have individual eyepiece focuses on them which once focused are in focus for any range, from a few dozen yards to infinity. These are my "ready use" glasses that I carry in my Jeep or leave on my desk. Each eyepiece is focused independently of the other and once focused is left that way. This is another type of focusing for binoculars, and frankly,

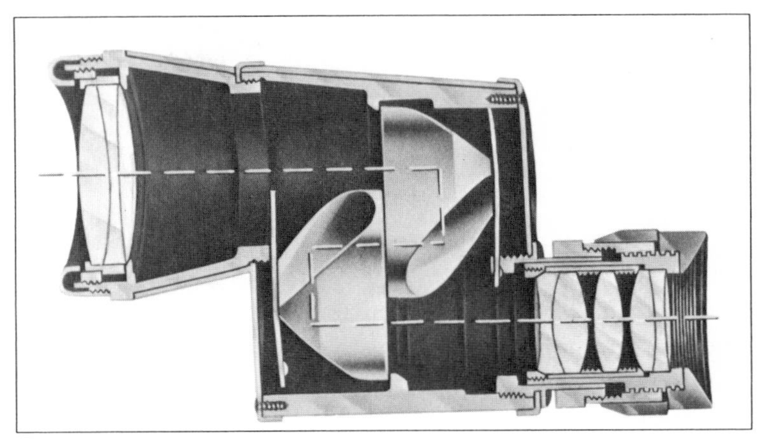

Cutaway of Bushnell Sportview 7×35 Wide-Angle Binocular. (Photo by Bushnell)

while such instruments cannot be passed back and forth between two users (unless the users' eyesight is compatible), they have proven their worth on many an occasion when I didn't really have time to mess around with focusing.

Eye Relief in Binoculars

Eye relief is critical in two optical instruments: binoculars and shooting scopes. In shooting scopes eye relief is as much a safety consideration as a use requirement. In binoculars it can determine how much you use the instrument.

When considering eye relief in binoculars you will quickly discover that not every firm in the business of selling binoculars really cares about the fact that a lot of their potential customers wear glasses. A long eye relief in binoculars is important because people who do wear glasses frequently cannot use binoculars without first removing their glasses. A good binocular should have an eye relief of around 9 mm and no more than 13 mm. At this distance most users who wear glasses can still make use of the binocular without removing their glasses. If you have to hold the binocular right up on the eye where it is uncomfortable to look through, then pass them up and find another brand.

Power in Binoculars

For some strange reason manufacturers are willing to sell American consumers anything they think they will have a use for. This includes useless binoculars. The absolute top power binocular I am capable of holding in my hand, without any sort of support, and using effectively is 8×. Beyond that power I must use a support of some type, whether it is my knees while sitting, or resting my elbows on the hood of my Jeep. I use 10× binoculars in my own outdoor observing but I don't carry them around with me when I am on foot.

The power range of binoculars is staggering. Recently I read an advertisement for a 15×60 binocular that sold for under $200. Bushnell is offering a 20×80 instrument; fortunately it is equipped with a tripod mount and adapter. I am firmly convinced that any object so far away that it requires any magnification over 10× to view should be seen through a telescope just so the vibration can be held to a minimum. The advantage of the very large binoculars is obviously in the stereoscopic vision area. A prismatic telescope would be a whole lot more compact and economical. Keep in mind that on a 15× or 20× instrument you are really talking about two telescopes, and there is no way that a 20×80 binocular is going to be built that can give you the image you want for the same price as a 20×60 telescope. To get quality in an instrument of that size you will have to pay for it. Bushnell's 20×80 binocular carries a price tag of over $500 and I would seriously question the quality of any binocular of that size which sold for less.

So what do you go for in a binocular? By now you should

have decided that I am a firm advocate of having two binoculars *and* a spotting scope, if you are serious about wildlife observing. A good binocular with 7×, 8×, or 10× coupled to a 50 mm objective lens provides maximum distance viewing of terrain and wildlife; and a good compact instrument of 5× to 7×, with a 25 mm to 35 mm objective lens, which fits in the pocket, insures that you can travel light on foot and still bring an object two or three hundred yards away into good view.

There are other instruments and one of the most popular is the *zoom* binoculars. I've used zoom binoculars and like them. When we are camping and want to just observe the wildlife around us, or I am studying the surface of a lake for feeding activity, zoom binoculars are great. In the case of zoom binoculars you have to be careful that the zoom feature is not a gimmick to sell an inferior product. Zoom, or variable power, binoculars require additional gears and elements within the eyepiece to function and you cannot get a variable power instrument without paying for it. Frequently you will find that you have to refocus after changing power or that at one or both ends of the power range the image will not be clear. There are good zoom instruments on the market but these are not binoculars that are normally hauled around in the back of a Jeep bouncing on an old logging road. In most cases you are better off with two instruments rather than trying to take a short cut unless you have the paycheck which allows you to invest in a quality zoom instrument.

Choosing Binoculars

I have a friend who owns a moderately priced binocular he bought in a discount store. It took him two months to find the instrument he wanted at the price range he was willing to pay. You couldn't get that binocular away from him with a pry bar. I have another friend who comes around every hunting season

Bausch & Lomb Discoverer 7×24 Compact Rubber Armored Fogproof/ Waterproof Center Focus Binocular, Model 61-7242. (Photo by Bushnell)

wanting to borrow one of my Bushnell Discoverer compacts. Last year I let him borrow one — for the first, last, and only time. They now sit on my shelf useless. I am sure that this year he will want to borrow the other pair and the answer is simply "no." I use my binoculars more than anyone I know, from simple night sky observing to serious field work. How you choose binoculars or any other optical instrument must be determined by your use. If you are going to sit in your kitchen looking at sparrows feeding on your lawn, you don't need an armored, fogproof instrument. If you are headed into the back country to study the feeding habits of the bighorn sheep, you need an instrument that is built to stand up to the abuse.

Before you go to buy binoculars determine the use you are going to put it through and follow that up with how much you plan to pay. Don't buy a pair of binoculars because it looks "macho" or because you like the case or strap. Test each and every pair. Look for the faults I've covered and check for tightness. Examine the lenses, the exit pupil. You don't have to spend your month's pay to find what you want, but you do have to put some effort into it. When you buy your binoculars be sure you are buying what you want and not what the salesman wants you to buy.

One final note on selecting binoculars. Never buy the ones that have been sitting in the display case no matter how good the bargain may seem; remember, people who don't own them don't care. For the same reason, when you buy *any* optical instrument never loan it out. You will never get back the same instrument you let go out the door.

9
Observing Through Microscopes

I have a friend who visits me on occasion in southern Colorado and his most frequent comment is he is never sure if he is in the office of a mad scientist or outdoor writer. Scattered around the nooks, crannies, bookshelves, and closet of my office are telescopes, binoculars, microscopes, and the usual amount of additional clutter created by the supporting paraphernalia.

Microscopes may seem out of place in a book about observing wildlife. I don't think so. Nature is the sum of our natural world. Our world extends from the heavens to the smallest unseen part of the world. On our camping trips I've collected samples of pollen, fur, plants, even stream water, and just about anything else which interests me. I carry small glass specimen bottles in my fishing vest and cart home samples to become part of my collection of microscope slides. Looking into the microscopic world can be as fascinating as watching the bighorns on a mountain side.

Microscopes are also educational. I've collected water samples from "clear" streams in the mountains around our home and let visitors see the parasites living in the "clear" water.

A compound microscope.

As a expansion of your enjoyment of the natural world, and whether for your children's education or for your own curiosity, a microscope can be a good investment. There is the added bonus of having an instrument in the house that from time to time may answer a question. Depending on how involved you want to become you could find yourself with a new hobby.

The Compound Microscope

Although the microscope is a complex instrument to build, because it works with such small lenses, the basic operation of the instrument is quite simple in principle and not unlike the telescope. The difference is that a microscope magnifies the object hundreds, or even thousands, of times. The compound microscope covered in this chapter is the most common microscope and the one available to the consumer. There are different types of microscopes, for specialized types of viewing, that are used by researchers but these are outside our area of interest.

Parts of the Microscope

The base of a good instrument is made from a heavy material that will hold the instrument in place when the microscope body is tilted toward the user for easy viewing. All of the microscope will tilt on this base and moves as a unit. The mirror for collecting light is the lowest part of the body and is mounted so that it can be adjusted to focus light through the microscope stage. Most manufacturers today offer microscopes with a two-way light source. One side is a mirror for reflecting natural light and the other is a small light for when a reliable light source is not available.

Directly above the mirror is the stage. This is the flat platform with a hole in the center and two metal clips for holding the slides in place. The stage may be rigidly mounted to the arm of the instrument or, in other models, is mounted on a

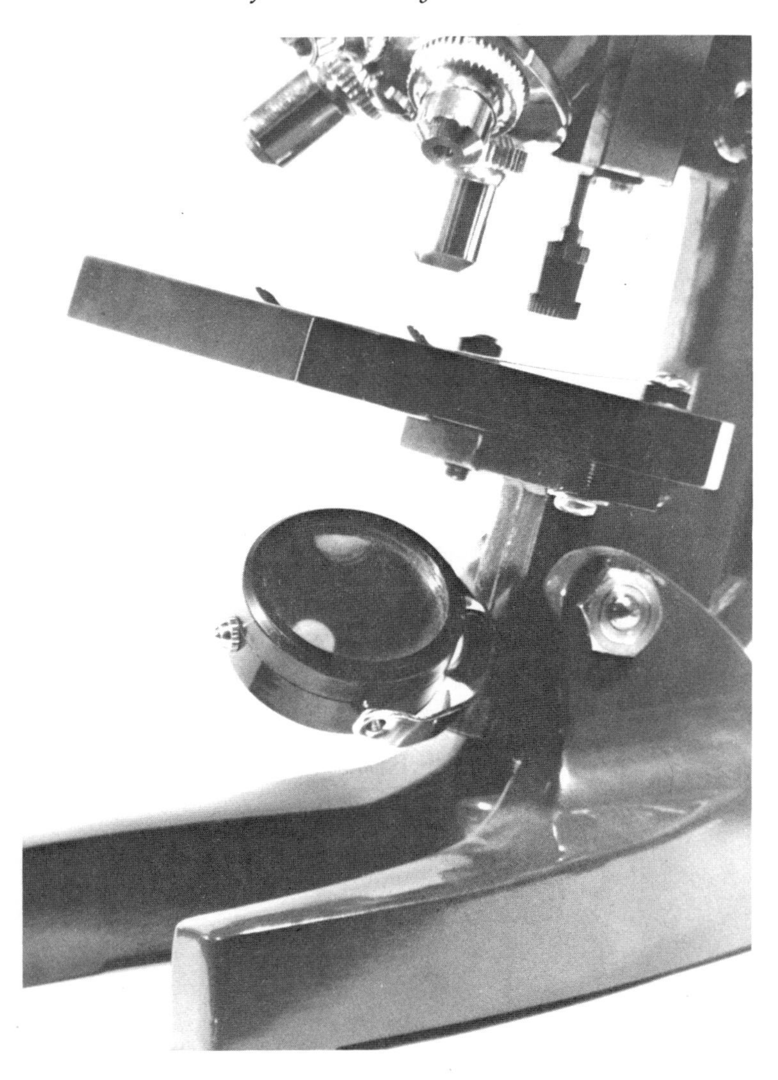

Mirror for reflecting light through stage and slide to the objective lens. This objective lens turret has four different lenses. Note the focus stop set screw.

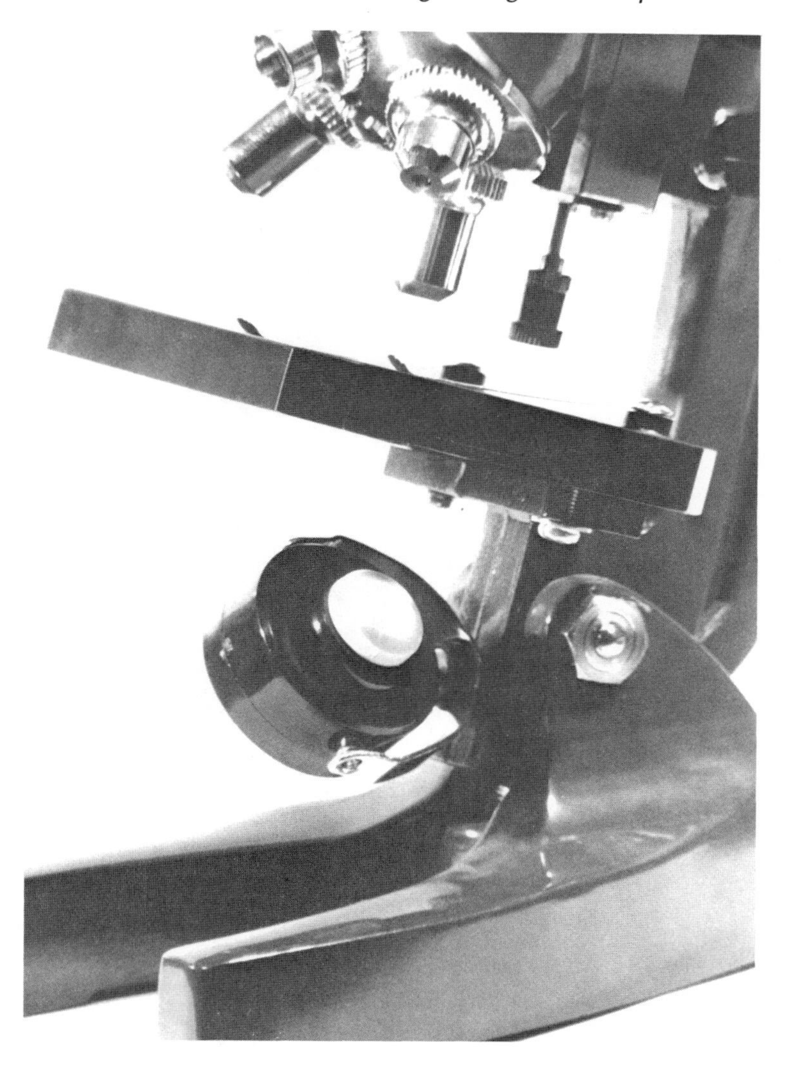

Same view of microscope only the mirror has been turned over to show the electric light source.

Eyepiece of microscope. The zoom power is set on 20×. The large knobs are the main focus and the critical focus knobs are below.

set of fine gears and is adjusted as the fine focus phase. Above the stage, at the bottom of the tube is the turret and objective lenses. These objective lenses produce the initial magnification of whatever is being viewed. Today's microscopes have from three to five objective lenses mounted on a moveable turret with magnification ranging from 4× to 60×. These objective lenses form a real magnified image within the barrel of the microscope which is then picked up by the eyepiece lens and is again magnified for viewing. In some instruments the eyepiece lens is a zoom type, or you can obtain interchangeable eyepieces to change the magnification. As I explained for binoculars with zoom lenses, these systems require additional lenses and precise gearing. Interchangeable eyepieces are usually superior to the zoom eyepieces in lower-priced instruments.

Whether zoom, fixed, or interchangeable the actual power of magnification in a microscope is arrived at by multiplying the power of the objective lens by the power of the eyepiece. As an example:

objective power is 12× and eyepiece power is 10×;
12 × 10=120× final magnification.

A microscope equipped with a five-way objective lens turret and two or three interchangeable eyepieces will offer from ten to thirty different magnification possibilities.

Most commercial microscopes are available with magnification powers ranging from a total magnification of 40× up to 1200×. Most of my viewing is usually within the lower ranges of magnification. The higher powers are handy from time to time to observe the cellular structure of specimens; however I always start with the lowest possible combination and slowly work my way up the magnification scale until the power is empty magnification and I am unable to make out what I am viewing. At

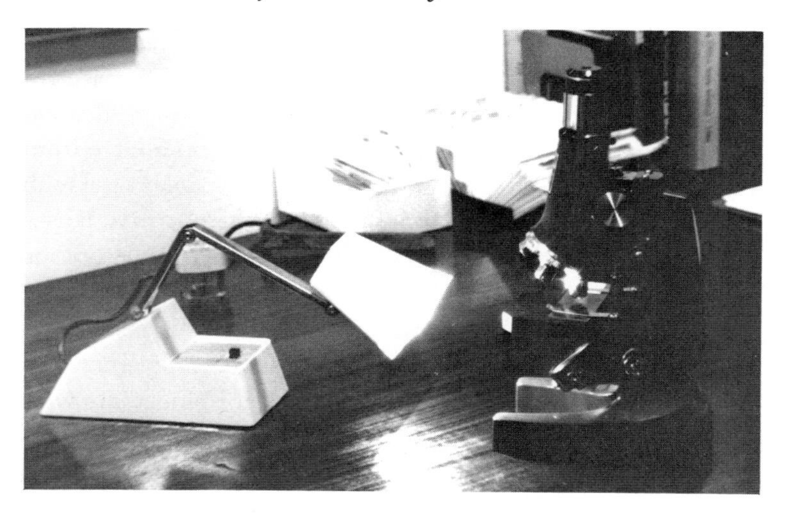

A high intensity desk lamp is an ideal light source for using the microscope mirror.

that point I back the power down until I have settled on the best power for viewing.

Whatever power you are using will require a very critical close focus to obtain a clear image. Most instruments have two focus knobs. The first is a coarse focus, which is the larger knob highest on the body of the scope. The second is the fine or critical focus, which is a smaller knob located below the coarse focus.

One microscope I own has a color filter wheel between the mirror and stage. Color filters allow you to alter the spectrum of light passing through the object being viewed. From time to time I use one of the filters rather than using dyes or stains on the slide.

A safety feature that is handy on a microscope is a "down stop." This is a simple set screw device that prevents the objective lens of the highest power from touching the microscope

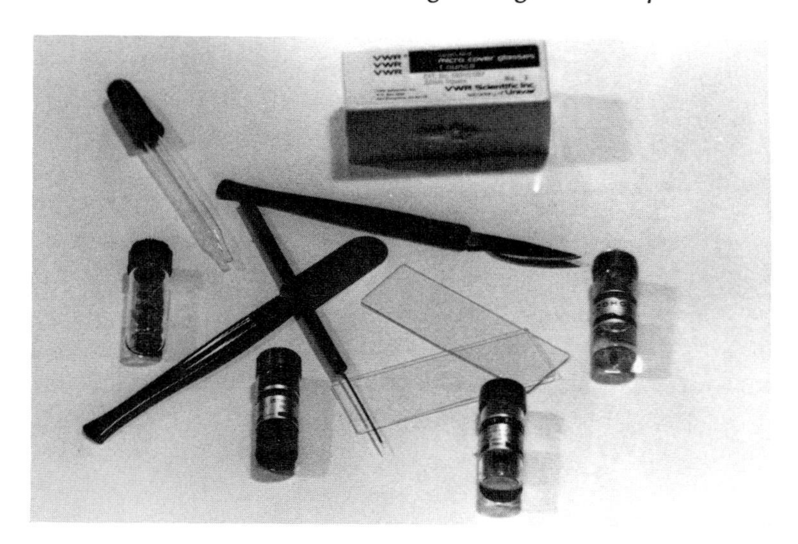

Extra tools needed include glass eyedropper, dissecting tools, alcohol, Canada balsam, stain, glass slides, and cover slips.

slide. Before viewing any slide on the highest power this screw should be set so the objective lens will not touch the microscope slide. Otherwise, when focusing, you can actually drive the larger objective lens through the slide.

Additional Tools for Microscope Use

If you acquire a microscope that does not have an electric light source then you should invest in a small desk type high intensity lamp that can be used as a reliable light source. The best source of light for microscope viewing, however, is sunlight, but it is not always available.

Most microscopes are sold as kits and will include a few prepared slides and some blank glass slides as well. Years ago I was able to walk to the local drug store and buy microscope slides, cover slips, dissecting tools, specimen bottles, and other

Specimen bottles are available in different styles and sizes.

science hobby tools. I doubt if you could find a drug store stocking any of those tools today. Today you need to locate a scientific or medical supply house and buy your slides and extra tools directly from them. In the state of Colorado I was able to locate only one supply house where I could buy a dozen glass slides and an ounce of cover slips. The tools you will want for your microscope observing include a good dissecting kit (which will make the job of preparing slides easier), a supply of glass slides (both flat and well slides), cover slips, Canada balsam, xylol, iodine, glass eyedropper, storage case for glass slides and another for samples collected in the field, and a specimen kit which carries a dozen small specimen bottles is handy.

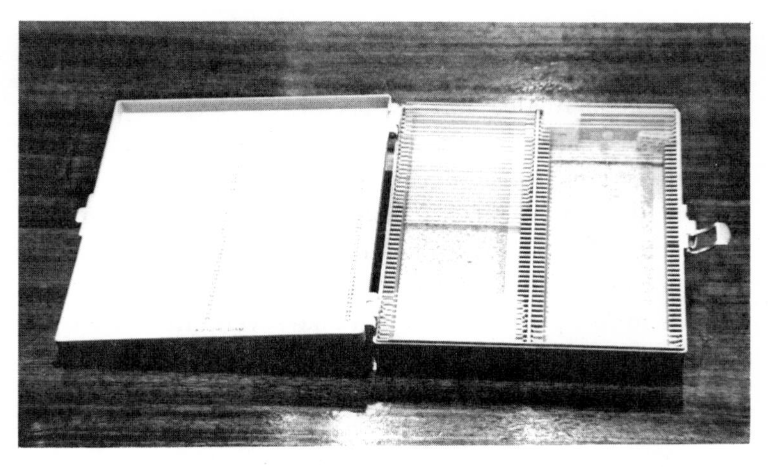

Microscope slides should be kept in a good storage case for safety.

Making Slides

I first started making microscope slides in grade school and, surprisingly, find it is a skill I have called on from time to time in the course of my life. When I checked with some of the youngsters around my neighborhood I was dismayed to learn none of them had been taught how to make a microscope slide. If you don't know how to make a microscope slide then a microscope is useless.

There are two types of microscope slides—dry slides and wet slides. Dry slides are used for mounting feathers, fur, plant parts, or other specimens that will not decompose, while wet slides are used to mount parts of insects and similar items. Both wet and dry slides can be either permanent or temporary mounts.

To make either type of slide you must first prepare the

specimen. Some specimens such as pollen do not need any preparation because they are already small enough for mounting. Others, such as leaves, flower petals, etc., must be prepared. In order for light to pass through the specimen it must be sliced in thin sections. Most specimens can be cut in either cross sections or vertical slices. I usually make two slides of any specimen, one cut each way. To cut off a section for viewing your slice or cross section must be as thin or thinner than paper. Practice makes perfect and it will take several tries and some patience before you are able to make the thin cuts consistently. Your blade edge must be razor sharp without a ragged edge. Hold the specimen firmly with one hand and cut in one smooth stroke so the slice you want to mount is nearly transparent. If you are making one or two slides of a specimen have clean slides and cover slips ready. Use tweezers or the point of a tiny paint brush to pick up the slice by the very edge and place it on the slide.

For cutting slide specimens of grass stems, evergreen needles, or other very hard, or very soft, materials place the specimen between two pieces of paraffin, then slice the paraffin and separate it from the specimen by placing the slice in a small amount of water. For fibrous material it is easier if a small piece of the material is placed in a drop of water on a slide then gently shredded with a dissecting needle.

Your slide should be clean and ready for the specimen before you cut. If you are cutting several slices from the same specimen it is easier to cut the slices and place them in a small dish of clean water until you are ready to mount them. Again, use tweezers or the paint brush to move the slices from the water to the waiting slides.

Before moving the sample from the water, or cutting it, prepare the slide by first cleaning it using lens tissue, then place a single drop of water in the center of the slide. Remove the specimen from the dish if you precut the samples, or make your slide then place the specimen on the drop of water. Take one of the cover slips, again using tweezers, and place it on its edge on the slide, then lay it down on the drop of water and specimen starting at one side so that the air bubbles are pushed out of the water. Once the slide is in place you can gently tap the cover slip glass with the blunt end of a dissecting needle to seat it. It will take a little practice to learn just how large the drop of water should be.

On some slides you may want to stain the specimen for better viewing of the cells and fibers of the material. A small drop of water colored with iodine can be placed on one side of the cover slip and a tissue held against the other side of the slip to draw the iodine through the water under the slide.

This temporary mount is now ready for viewing. I have had some temporary mounts remain intact for several weeks but eventually the cohesiveness of the water, and the vacuum formed when the water evaporated, will let go and the slide will fall apart.

One of my perfectly useless hobbies in outdoor observing is collecting fish scales and making slide mounts of the scales. I want these to be permanent mounts because I am frequently comparing the scales of fish of the same species from different waters. To make a permanent mount follow the same procedure for making a specimen, however this time place your prepared specimen in the center of the slide without the water. On top of the specimen place a drop of Canada balsam rather than water. Holding the cover slip lay it on the specimen then press down gently with your dissecting needle so the balsam spreads to the edges of the cover slip and pushes out the air bubbles. Place a weight over the slide and let it set for a full week before moving the slide. After the balsam is dry use a small label to date and identify the slide. If you have balsam on the edge of the cover slip use xylol to clean off the excess.

Wet slides are made the same way as dry slides except the specimen must first be placed in alcohol for about twenty minutes. This is a "fix" bath that will harden the cellular structure so it will not decompose. After it is fixed remove the specimen from the alcohol and place it in xylol for two or three minutes to remove the alcohol, then place the specimen on the slide and continue, using the balsam as you did for dry slides.

Ground Well Slides

A recent article I wrote was the product of spending over a hundred dollars for veterinarian care for my springer spaniel who had picked up a small grass seed that worked its way into her leg and migrated up the muscle, finally erupting several inches up her leg. For that article I needed to photograph these seeds under a microscope. Because of the size of the seeds I could not make microscope slides using flat glass slides but had to use *ground well slides*. These are slides which have a well or depression ground into them allowing you to mount larger specimens, such as grass seeds. Specimens are mounted on these slides in the same manner as for flat slides, however more care must be taken to prevent air bubbles.

Photography With Microscopes

As a kid I wanted a microscope kit with a camera attachment. I did not get one until a few years ago and since then I've enjoyed photographing different slides and building a small collection of photographs of hobby projects. In addition I have used some of the photographs to illustrate articles I've written.

The photographic attachment for a microscope is really a simple device and today most kits that sell the attachment offer

Photo of group viewing and photographic attachment on microscope. Attachment replaces the eyepiece.

it as a part of a group viewer hood, which eliminates the need for staring through the eyepiece. Most of these attachments use 126 cartridge film. At one time this film was available at any store but recent camera changes have begun to push this film out of the market and in a few years it may no longer be available. If you find a store that has a supply of this film buy enough to last you for several years if you have a camera attachment.

Microscope viewing adds another dimension to wildlife observing. The bird watcher may find it fascinating to collect the feathers dropped by different birds and build a library of slides of the feathers. Or, you may find yourself fascinated by the types of pollen in wildflowers and begin cataloging them. Whatever direction you take keep your slides in a microscope slide case, label them, and keep a notebook recording what each slide is, when and where you collected it, and any other information which will help you compare the different slides. Like using the telescope or binocular, microscope observing is just another way of getting closer to our natural world.

10
Observing — Not *Intruding*

Wildlife biologist Timms Fowler and I agreed to meet at a small truck stop on the edge of Lamar, Colorado, at 4:00 A.M. After breakfast and a review of the schedule we piled my gear into his pickup and drove south of Lamar to spend the day observing bald eagles. I was covering the story as a freelance assignment from the *Colorado Springs Sun* and was looking forward to it. Timms was studying the habits of bald eagles living near and around Two Buttes, Colorado.

We parked the truck two miles from where he intended to do his observing, strapped on our packs, and walked across the snow-covered plains to a distant rise. As we neared the crest we dropped to our hands and knees and crawled through the snow-covered cactus to a fence. There he opened his pack and took out his spotting scope, tripod, notebook, and thermos of coffee. Throughout the morning we lay in the snow while he studied the eagles. Every time the eagles moved, whether it was flexing their wings or taking off to make short circles around the cottonwood tree roost, he made notes. Later, when the eagles left their roost to hunt near the reservoir, we walked down to the trees to collect dropping samples, hoping to find out what the eagles had carried to the roost and eaten.

Every day Timms Fowler observed the eagles. His biggest fear was that he would get too close and frighten the birds from their roost. To the best of my knowledge the majestic birds never knew he was watching them because he kept hidden and maintained a careful distance so they never felt threatened.

It can be argued that man is part of the natural world; and it can be argued that man is not. Whichever side a person takes is a moot point. What is important to understand is that all wildlife is easily disturbed by humans. Whether a person is walking along a trail or hidden in a blind, the human presence will have some effect on the wildlife. To be a successful observer of wildlife, whether for recreation or as a researcher, the greater the distance you can put between yourself and the animals you are observing the less impact you will have. Sometimes this is not practical and you must be close to the wildlife you are observing. To get close requires planning and patience.

There is another element that must be considered and that is that *all* wildlife species possess the ability to present a danger to any observer. On one occasion, eagles, frustrated by a low-flying, light airplane being used to count their numbers along a canyon, attacked the plane, causing it to crash killing both the pilot and the researcher. Wildlife photographers have been treed by bears, mauled by lions, and bitten by snakes. There is an element of danger in wildlife observing that increases as the distance between the observer and the wildlife is reduced. There are six rules for wildlife observing which can keep you out of trouble.

1. Never approach a wild animal's young.
2. Never place yourself between a mother and her young.
3. Never disturb a nest, pick up a young animal, or use food as a bait to draw animals to you.

Sometimes a blind for wildlife observing needs to be nothing more than something "non-threatening." Dave Colwell and author spent an afternoon watching pronghorn while sitting beside a Jeep.

4. Always maintain the greatest possible distance between yourself and the wildlife you are observing.
5. Always observe wildlife with another person acting as a second pair of eyes watching around and behind you.
6. Always leave yourself an escape route.

These rules are simple but seldom followed. I was in eastern Colorado trying to photograph snow geese when the birds suddenly exploded from the lake. A little while later I ran into a "wildlife photographer" who was frustrated because the birds had taken off as he approached the lake shore. On another occasion I was walking toward a favorite beaver pond for an afternoon of fishing when I found a young woman trying to get a newborn fawn to hold up its head and look at her camera. I convinced her to leave the fawn alone. Two days later I was returning to the lake and found the fawn had died. Its mother abandoned it after it was handled by a human.

The most important rule of wildlife observing is not to intrude. Watch, observe, photograph, but always from a distance. There are two acceptable methods of wildlife observing. The first is stationary observing from a prepared blind and the second is mobile observing. Each one requires a little effort.

Stationary Observing

Timms Fowler did not use a prepared blind to observe the eagles while working for the Bureau of Land Management on the eagle project. But he always watched from the same pile of weeds along the fence. He had scouted the area thoroughly and found a natural windbreak where tumbleweeds piled against the fence and he capitalized on that, using the natural pile of weeds as his blind. There are times when it helps to build a blind to observe wildlife. Commonly, blinds are built because the birds or animals to be observed will either return to the area in the spring or fall and spend several months there, or the animals live in the area throughout the year and the blind will allow you to observe them in their daily yearround lives.

Determine if you need a blind by what you are going to be doing. As an example, if you know of a pond where ducks nest and raise their broods every year and you want to observe and photograph the entire nesting season, you should build the blind in the winter, well before spring arrives, and use plants from around the pond to camouflage the blind. If possible transplant cattails or other plants from the pond to the area around the blind; their natural growth will help camouflage the blind.

The blind should be comfortable so you can spend long periods of time inside. Professional observers have learned that shuffling noises will startle the birds and animals around a blind

A blind dug into the side of a hill for observing a water hole. Note the transplanted clumps of grass around the blind.

so they cover all the exposed wood or other surfaces with carpet or foam rubber. The entrance to the blind must be hidden and securely closed when you are not inside. More than one person who built a blind has returned after several days to find skunks claiming it as their new home. The access route to the blind must also be hidden so that you can slip in and out without disturbing the wildlife in the area.

If your plans are to use the same blind throughout a season build it from wood, making the sides, top, and bottom solid. Your observation windows should be protected with hinged shutters that you can open and close from the inside. Cover the windows with fine screen netting that can be opened and closed to allow you to push a camera or telescope lens through. The extra effort will keep troublesome insects out. Build a small work space shelf below the windows so you don't need to pull a table around, which might make noise. Furnish the inside with a comfortable chair.

Outside the blind cover the sides with chicken wire then weave grass and leaves into the chicken wire using only plants collected from the vicinity. Transplant living plants around the blind to add to the concealment. Finally, well before the arrival of the first wildlife in the spring, leave several articles of clothing you frequently wear inside the blind. The animals will smell the clothing, become accustomed to it and accept your smell.

If your blind is on private property, or property for which you can control the access, then you can add a hasp and lock the blind when you leave so that you can leave your observing tools there through the season. Leaving the telescope or camera equipment in the blind will make it easier to slip in and out without disturbing wildlife.

If your blind is on private land all you need is the landowner's permission to build the blind and to abide by whatever rules the landowner puts down. However, if you plan to do your observing on public lands, whether they are federal, state, or municipal, you must obtain permission from the responsible agency. In some cases a permit will be required and you will have to provide a good reason for building a semipermanent structure on public lands. If your blind is on public lands you will also run the risk of vandals destroying the blind or stealing your equipment.

When the season is over you must decide if you are going to maintain the blind or dismantle it. If you are not going to use it the following season then dismantle it and remove as many traces as possible of the blind's existence. If you are going to use it again the following year then do your repairs, modifications, or other maintenance between observing seasons. Once the wildlife you intend to observe arrive your blind will remain effective only as long as you stay in it when in the area. Getting out to make repairs will upset the surrounding animals and they may drift away from the area.

If you set up a blind for observing an entire season you will see nature as it is. If you are observing ducks nesting, at some point you will see a fox or coyote sneaking up to snack on the ducklings. The natural urge is to warn the ducks. Don't! Nature has provided the brood with the numbers and the tools to survive. The fox and coyote must live as well. Just observe nature and don't become a participant. It is hard to watch a duckling stolen from its family, but it is nature at work.

Mobile Observing

Mobile and casual observing is where most wildlife observers get into trouble or cause problems. If you want to hike into the mountains to observe the feeding habits of the bighorn sheep be sure you are physically able to undertake the hike. Once you are in the mountains keep a good distance between you and the wildlife you are observing. I've watched a herd of animals for an entire afternoon by keeping distance between them and myself. Once a friend and I pulled a Jeep to a ridge top, put up a sun shade, got out our telescopes, and spent the day watching a herd of pronghorn that ranged from more than a mile to less than fifty yards from us. We did nothing threatening toward the herd and they ignored us. If you are hiking and see a herd of animals don't try to sneak up on them by leaving the trail. Oddly enough most animals are well aware of the trail's existence and that humans use it frequently. A person on a trail, even if they have stopped to watch deer feed, present no threat. Off the trail however, the human is a threat and the deer will run. If you want to observe wildlife don't threaten it. One winter I was in the mountains with a friend and his wife and we were

Wildlife observing is sometimes frustrating. The two ducklings are the last survivors of a brood of six; the others were picked off by predators.

watching the remaining finches and a few other birds in the snow. After half-an-hour of quiet watching the birds had accepted us. I held out a gloved hand and a finch landed on it, sat for several minutes, then fluttered to a nearby limb.

You will observe more wildlife by moving slowly through an area and stopping frequently. Camouflage clothing, face net, and hat are great assets. Find an area that has the needed foods and water for wildlife and rather than walking through to see what birds are in the area move into it slowly, then sit down in the shadows to observe. You will be surprised at how quickly birds and small mammals accept you as a part of the landscape. Your telescope and large binoculars are useless here, although a pair of folding roof prism binoculars will become invaluable. You can even make notes while you sit. You may discover a nest in the area and you can come back later to observe from a safe distance with your telescope.

After spending an hour or two in one spot you can move a hundred yards or so to another area and sit down again. Because most birds and small mammals are very territorial you will see a different collection of wildlife. If your move of a few hundred yards brings you from a clearing to the edge of a natural spring that is surrounded by pine trees, the wildlife around you will change as well.

Mobile observing allows you to move through an area and to learn what species are found in it. Later you can locate hiding places from which you can safely observe the animals living out their daily lives. Don't try to carry a spotting scope or heavy binoculars when moving. Return later to use the larger instruments and always do your observing from a distance.

The Ethics Questions

Wildlife observers are not participants in what is taking place around them. If you see a predator stalking a grouse, it is their world and you are observing nature provide food for a species. Occasionally you may find an injured animal. As a youngster I once found a hawk that had broken a wing and asked my father what to do. He called the Oklahoma Division of Natural Resources who came out and trapped the hawk then delivered it to a vet who specialized in caring for wild animals. That should be the closest you get to interfering in the wildlife around you. Every state has a specialist who can trap an injured animal and provide it with care and who will ultimately return it to the wild, unless its injuries are so severe that it can never fend for itself again, in which case they will usually transfer it to a zoo or even keep it as part of a nature program.

You may be observing an area to suddenly discover campers or hikers invading the area. Public lands are public lands and as an observer, unless you are under contract to the agency controlling the land, you cannot interfere with others' enjoyment of the area. The same is true during hunting season. Regardless of your personal position on hunting the hunters are there and as an observer you should not interfere with their hunting. The wildlife observer is on the outside looking in, not participating. However, as an observer, if you ever see a wildlife law violated by a hunter or anyone else, report the violation to the nearest law enforcement agency.

Observing the Rare Species

Over the years many observers have been frustrated because they have spotted a rare species in an area and after reporting it to the proper agency were dismayed to discover a "ho-hum" attitude. The problem is that most wildlife managers would like to believe the sighting but don't have the evidence they need. Consider the day my wife and I observed the Canadian lynx. At first we were frustrated by the wildlife official's denials of the lynx. We learned later they knew the cats were in the area but were keeping it secret. When we were able to give a report in private every detail was taken down in serious enthusiasm. Keep in mind that rare or endangered species are important to wildlife managers and they want information you might have, but not publicly! In fact, if you get pictures of an endangered species you may be asked not to talk about your sighting for fear of attracting attention to the area. On the other hand, if you are able to provide detailed information from photographs or other hard evidence they will welcome it. Once I collected deer droppings from an area and turned them over to the area wildlife manager who sent them to a research center studying the effects of deer living around orchard areas. I have also collected eagle casts and sent them to a friend studying the feeding habits of eagles on a nearby reservoir. On another occasion my wife and I spotted an osprey and jotted down the exact location, time of day, and the direction the bird took when it flew off. When we got home that afternoon I called the Division of Wildlife to report the sighting. I was transferred to a researcher who was directing a project to study the birds. She was elated with the news and gave me a number to call the next time we spotted an osprey. Later that week my wife and I saw a pair near the same area and I called the number and gave her the information. Several weeks later I received a letter from her explaining that the information my wife and I passed on provided her with a reason to conduct a search of the area, during which she discovered where the birds were nesting. The area was quickly closed to the public and the pair nested in private, raised their young, and are part of a continuing research program. The lesson is clear: always carry a notebook and make exact notes of your observations. You may play a minor or major role in providing information needed by researchers.

Whether you build a blind, don camouflage clothing and walk through an area, or just drive along country roads watching the wildlife, remember that you are an observer. Your optical equipment should be the aids that provide a closer and clearer window into the world of nature.

11
Take Care of Your Optics

Although this chapter deals with how to take care of the optical equipment you own, I should also mention that it is important to take care of the optics which belong to other people. This is true whether they have loaned the equipment to you or you are simply hiking together. I have had equipment damaged by friends who dropped their gear on top of my telescope, dropped the equipment in water, left it outside overnight, cleaned lenses with tee-shirts and spit, and generally abused it.

I have already pointed out that the idea of fogproof optics eliminates the need for worry about dust getting into the system and in fact makes the system waterproof as well. Modern equipment is also rugged—not so rugged, however, that it can be mistreated. I have made the mistake of loaning out some of my equipment to others only to have it returned to me ruined. Sitting on my shelf is a very expensive Bausch & Lomb compact binocular that was ruined by a friend who promised to take good care of the equipment.

If you want to ruin a good optical instrument of any type leave it in your car or pickup where it can get very hot. The cement holding the lens elements and other small pieces of glass together becomes soft in extreme temperatures and the constant shuffling from one extreme to another weakens and finally destroys the system. If you do have a scope or binocular that you keep in your vehicle, place them in an area where they will be exposed neither to direct sun or high temperatures. Just about anything can be ruined by the intense light coming through a car window and the inside of an automobile left in the sun can soar to over 140 degrees F in less than half-an-hour. As a final note, carry any optical instrument in the case designed for it, or buy a case that will protect it. When the equipment is not in use keep it in the case. If you live in a dusty area take the extra precaution of draping plastic over the case. Don't wrap the instrument itself in plastic; condensation will build up and damage it.

It should also go without saying that both the objective and eyepiece lenses on an instrument should be stored covered with lens caps. It seems like a simple thing to do yet it is one of the most difficult for people to remember. The lens cap keeps the dust and moisture off the exposed glass.

After a day's observing, before putting your equipment away, let the system air out by removing the lens caps so that any moisture that might have collected in the space between the lens cap and the lens will evaporate. Do this with *all* of your equipment, letting it air out for 15 minutes even if it is fogproof. After it has aired out use a very fine camel hair brush with an air blower attachment to blow and brush away dust and debris that has collected on the instrument. After it has all been blown away wipe the metal parts with a soft rag, and, finally, use a clean lens rag to wipe down the glass. Do not use tissues or napkins to clean a lens and if there are no water spots on the lens the camel hair brush is sufficient. If you do have dried water spots then use photographic lens cleaner or contact lens cleaner very lightly to clean the lens. Never use commercial glass cleaner that is sold

Clean your optical gear after every outing with a soft air brush and lens tissues.

for washing windows, water, soap, or spit to clean a lens. Do not use so much of the proper cleaner that it can seep into the area around the lens. Although the lens is fogproof some types of lens cleaners can eat at the cement around the lens.

If you are using your equipment at night, especially if it is a telescope or spotting scope, never put it away when you call it quits. If you do have to put it in the case to get home, as soon as you are home take the entire system out of the case and set it up for an hour to let the night air evaporate. If you can leave it up for the entire night in your living room or cabin, then you are even better off.

If you are in a position where you are going to be using your telescope a lot at night or in foul weather, then a good investment is a dew cap that goes over the front of the instrument and extends beyond the objective lens and prevents most night

After any observing session it is a good idea to let the equipment "air out" before putting it away.

condensation from collecting on the lens. A dew cap will also help protect the objective lens from snow and other forms of falling moisture.

Finally, when you put your equipment away for extended periods be sure you have removed all of the dust and moisture and wiped the instrument down so it is clean. Store your optical equipment where the daily dust settling on everything is kept under control. On gear that uses a battery for illumination or to power any part of the system remove the batteries so they cannot rust the system's electronics.

Most modern optical equipment is well built and can stand up to the punishment of the outdoors. A few precautions when bouncing around in a Jeep, such as not laying a scope or binocular on a hard surface, keeping all instruments out of sunlight filtered by the windows, and wiping equipment down every night, will insure years of service. On occasion I meet someone in the woods using the same equipment they have owned for 25 or 30 years and the gear is in excellent shape. It has always fascinated me that these same people are the traditional outdoor types.

There really isn't all that much to caring for optical equipment beyond a little common sense. There is more to its use and selection. Knowing what you are buying and why is more important than whether it "looks good." Taking care of the equipment, knowing the limits of the equipment, and investing in it properly from the start, then properly using the equipment for its intended purpose, makes the investment of a paycheck worth it. In today's world there is no shortcut to success, whether you are a professional wildlife researcher or an amateur observer. In every case attention to detail, an understanding of the equipment and dedication to getting the most use of the equipment is the only route to success. Anything else is a myth.

Index